Table of Contents

Introduction

Introduction

Introduction

WTF?! What's the factor? I ask my students this all the time. And many times, I get the WTF?! confused look back from them.

The goal of this book is to provide a basic understanding of one of the most essential concepts and skills in algebra, which is factoring. Through all my years of teaching all levels of math from middle school, high school, and college, I have found that many students have difficulty with this at an early level, and thus makes it harder for students to be successful in later math courses. This book is an expanded form of my lecture notes and includes extra explanations, examples, and practice. I will be focusing on solving as many problems by hand, but also showing how graphing calculators like Desmos can be used as well. Solutions to practice sets are at the back of the book. This book will start off with the very basics of arithmetic and dig deeper into the concept of factoring as it is used in algebra and precalculus.

Multiplying Numbers

Multiplication is a mathematical operation that combines two or more numbers to give a single result, called the product. The product is obtained by adding one number to itself repeatedly as many times as the other number specifies.

For example, consider the multiplication of 3 and 4. To find the product of 3 and 4, we add 3 to itself four times: 3 + 3 + 3 + 3 = 12. So, the product of 3 and 4 is 12. We can write this as: 3 x 4 = 12. The "x" symbol represents multiplication in this case. We can also write multiplication using parentheses: (3)(4) = 12 or with a dot $3 \cdot 4 = 12$

Multiplying Numbers

Multiplication can be used to find the total quantity of items when we know the number of groups and the number of items in each group. For example, if we know that there are 5 groups, and each group has 3 apples, we can find the total number of apples by multiplying 5 and 3: 5 x 3 = 15. So, the total number of apples is 15.

Multiplication can be represented visually or geometrically as a rectangular array. To visualize multiplying 6 by 4, imagine a rectangle that is 6 units long and 4 units wide. The area of the rectangle would be 6 x 4 = 24. If we divide the rectangle up by unit squares, the rectangle has 6 squares across and 4 squares down, giving 24 squares total. It can be visualized as such:

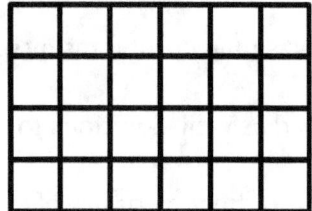

Multiplication has several properties including the commutative property, associative property, and distributive property.

Multiplication is commutative, which means that the order of the numbers does not affect the result. For example, 3 x 4 gives the same result as 4 x 3:

3 x 4 = 12

4 x 3 = 12

Multiplication is also associative, which means that the grouping of the numbers does not affect the result. For example, (3 x 4) x 2 gives the same result as 3 x (4 x 2):

(3 x 4) x 2 = 12 x 2 = 24

3 x (4 x 2) = 3 x 8 = 24

Multiplication also follows the distributive property that relates multiplication and addition. It states that when we multiply a number by a sum of two or more numbers, we can distribute the multiplication over each addend, and then add the products. Mathematically, the distributive property can be expressed as: a x (b + c) = a x b + a x c

In this equation, a, b, and c are any real numbers or algebraic expressions.

The distributive property can also be extended to subtraction, which gives us:

a x (b - c) = a x b - a x c

To understand the distributive property, let's consider an example: 2 x (3 + 4)

Using the distributive property, we can distribute the multiplication of 2 over each addend, and then add the products: 2 x (3 + 4) = 2 x 3 + 2 x 4 = 6 + 8 = 14.

The distributive property is a powerful tool that allows us to simplify expressions by breaking them down into simpler terms. It is commonly used in algebraic expressions, where we use it to expand and simplify expressions. For example: 3(x + 2) = 3x + 6

NAME	DEFINITION	EXAMPLES
Commutative	You can change the order in addition and multiplication	$2 + 3 = 3 + 2$ $2 + x = x + 2$ $2 \cdot 3 = 3 \cdot 2$ $x \cdot 3 = 3x$
Associative	You can change the grouping (parentheses) in addition and multiplication	$(2 + 3) + 4 = 2 + (3 + 4)$ $(x + 3) + 4 = x + (3 + 4)$ $(2 \cdot 3)4 = 2(3 \cdot 4)$ $2(3x) = (2 \cdot 3)x$
Distributive	You can multiply a sum by multiplying each addend separately and then add the products.	$2(3 + 4) = 2 \cdot 3 + 2 \cdot 4$ $2(x + 4) = 2x + 2 \cdot 4$
Identity	When you add zero it doesn't change. When you multiply by one it doesn't change	$3 + 0 = 3$ $y + 0 = y$ $1 \cdot 5 = 5$ $1n = n$
Inverse	When you add the opposite, you get zero. When you multiply by the reciprocal you get 1.	$3 + -3 = 0$ $z - z = 0$ $\frac{1}{5} \cdot 5 = 1$ $\frac{1}{n} n = 1$

Multiplying Numbers

It is important to develop "number sense" by performing simple mathematical operations mentally as much as possible instead of relying on a calculator or technology. Developing this number sense will help you understand algebra when the "numbers" become more abstract variables and expressions.

When multiplying a string of numbers together, we can see from the properties above that it is not necessary to work from left to right as we see the numbers. Instead, you should look for patterns of easy numbers to multiply together. Multiplying a number by 10 is easy because we simply add a zero to the end. And multiplying by 2 is usually easy to double or add the number to itself. For these reasons, it is best to try to form 10's and 2's and keep those products last.

Here's an example of multiplying 2 x 3 x 10 x 3 x 4.

$$2 \cdot 3 \cdot 10 \cdot 3 \cdot 4 \cdot 2$$

$$2 \cdot 3 \cdot 10 \cdot \overset{12}{\cancel{3 \cdot 4}} \cdot 2$$

$$2 \cdot \overset{36}{\cancel{3}} \cdot 10 \cdot \cancel{3 \cdot 4} \overset{\cancel{12}}{} \cdot 2$$

$$\overset{72}{\cancel{2}} \cdot \overset{\cancel{36}}{\cancel{3}} \cdot 10 \cdot \cancel{3 \cdot 4} \overset{\cancel{12}}{} \cdot 2$$

$$\overset{72}{\cancel{2}} \overset{\cancel{36}}{\cancel{3}} \cdot 10 \cdot \cancel{3 \cdot 4} \overset{\cancel{12}}{} \overset{144}{\cancel{2}}$$

$$\overset{72}{\cancel{2}} \overset{\cancel{36}}{\cancel{3}} \cdot \cancel{10} \cdot \cancel{3 \cdot 4} \overset{\cancel{12}}{} \overset{\overset{1440}{144}}{\cancel{2}}$$

Practice Multiplying Numbers

Find each product using mental math.

1) 2×9 **18**

2) 4×11

3) 13×10

4) 8×3

7) 7×6

8) 6×3

9) 12×8

10) 14×13

9) 12×8

10) 14×13

11) $(4)(4)(9)$

12) $(7)(10)(2)$

13) $(5)(10)(4)$

14) $(8)(4)(7)$

15) $(6)(5)(4)$

16) $(9)(6)(2)$

17) $(3)(3)(4)$

18) $(6)(7)(2)$

19) $(4)(10)(7)$

20) $(9)(5)(6)$

21) $7 \cdot 4 \cdot 3 \cdot 10$

22) $2 \cdot 4 \cdot 10 \cdot 7$

23) $4 \cdot 10 \cdot 6 \cdot 5$

24) $6 \cdot 2 \cdot 10 \cdot 7$

25) $7 \cdot 2 \cdot 10 \cdot 2$

26) $7 \cdot 2 \cdot 8 \cdot 4$

27) $2 \cdot 10 \cdot 4 \cdot 2$

28) $10 \cdot 6 \cdot 2 \cdot 3$

29) $6 \cdot 8 \cdot 4 \cdot 5$

30) $5 \cdot 4 \cdot 7 \cdot 10$

31) $2 \cdot 6 \cdot 4 \cdot 2 \cdot 5$

32) $5 \cdot 12 \cdot 2 \cdot 3 \cdot 2$

Divisibility Rules

Divisibility describes the relationship between two numbers where one number can be divided by another number without leaving a remainder. In other words, if one number is divisible by another number, then the first number can be divided exactly into equal parts by the second number.

For example, if we consider the numbers 12 and 4, we can say that 12 is divisible by 4 because we can divide 12 into 4 equal parts, each part being 3. Similarly, we can say that 15 is not divisible by 4 because if we divide 15 by 4, we will have a remainder of 3.

Numbers that are divisible by two are called even. Even numbers end in 2, 4, 6, 8, or 0. Numbers that are not divisible by two are called odd.

Example 1: Determine if 36 is divisible is divisible by 2, 3, 4, 5, 6, 7, 8, 9, or 10.

Solution: $36 \div 2 = 18$ so 36 is divisible by 2. $36 \div 3 = 12$ so, 36 is divisible by 3. $36 \div 4 = 9$ so, 36 is divisible by 4. $36 \div 5 = 7\ R\ 5$ So, 36 is not divisible by 5. Numbers that are divisible by 5 end in a 0 or 5. Since 36 is divisible by 2 and 3 so it is divisible by 6. $36 \div 6 = 6$. $36 \div 7 = 5\ R\ 1$. 36 is not divisible by 7. $36 \div 8 = 4\ R\ 4$. 36 is not divisible by 8. $36 \div 9 = 4$ So, 36 is divisible by 9. $36 \div 10 = 3\ R\ 6$, 36 is not divisible by 10. Numbers that are divisible by 10 end in 0. Therefore, 36 is divisible is divisible by 2, 3, 4, 6, and 9.

Divisibility rules give us shortcuts to determine what a number can be divided by without any remainder. Using the divisibility rules will make it easier to simplify and work with fractions.

	A number is divisible by...	Divisible	Not Divisible
2	The last digit is even (0,2,4,6,8)	2018	2019
3	The sum of the digits is divisible by 3	2019 2+0+1+9=12	2020
4	The last 2 digits are divisible by 4	2020	2021
5	The last digit is 0 or 5	2020	2019
6	Is even and is divisible by 3 (follows 2 rule and 3 rule)	2016 2+0+1+6=9	2019
7	Double the last digit and subtract it from a number made by the other digits. The result must be divisible by 7. (This rule is most difficult on this list)	2016 201-12=189 18-18=0	2017
8	The last three digits are divisible by 8	2024	2020
9	The sum of the digits is divisible by 9	2025 2+0+2+5=9	2019
10	The number ends in 0	2020	2021
11	Add and subtract digits in an alternating pattern (add digit, subtract next digit, add next digit, etc.). Then check if that answer is divisible by 11.	2035 2-0+3-5=0	2037
12	The number is divisible by both 3 and 4 (follows the 3 rule and 4 rule)	2028 2+0+2+8=12	2037 Divisible by 3 but not 4

Practice Divisibility

State whether each given number is divisible by 2, 3, 5, 6, 9, or 10. The first one is done for you.

1) 24 **2, 3, 6**

2) 21

3) 15

4) 18

5) 25

6) 12

7) 17

8) 23

9) 13

10) 16

11) 32

12) 49

13) 38

14) 29

15) 39

16) 37

17) 40

18) 46

19) 28

20) 30

21) 63

22) 68

23) 66

24) 96

25) 74

26) 91

27) 53

28) 92

29) 75

30) 65

Factors of Numbers in Arithmetic

Factors and multiples are two important mathematical concepts that are frequently encountered in arithmetic. While they are related to each other, they represent fundamentally different ideas.

A factor of a number is a whole number that divides the number evenly without leaving a remainder. For example, the factors of 12 are 1, 2, 3, 4, 6, and 12 because they divide 12 without leaving any remainder. Note that factors can only be whole numbers and that every number has at least two factors: 1 and itself. The list of factors will be a finite or limited list.

When finding factors of a number, it is best to work systematically, making a list and pairing off the factors as we find them. Start by dividing the given number by 1 and you will get the number itself. These are both factors. Then divide by 2. If the remainder is zero, then 2 and the quotient are both factors and list them both. If the remainder is not zero, then we go to the next number, dividing it by 3. If the remainder is zero, then 3 and its quotient are both factors and list them both. If the remainder is not zero, then we go the next number, dividing by 4. We repeat this process until we reach the "middle" where the factors start to repeat.

Example 1: Find the factors of 36.

Solution: If we want to find the factors of 36, start by listing and pairing off 1 and 36.

Find factors of 36:   1,   36

Then try dividing 36 by 2 which gives 18. List the pair 2 and 18 like this:

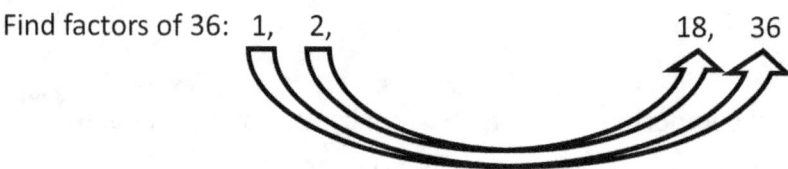

Then try dividing 36 by 3 which gives 12. List the pair 3 and 12 like this:

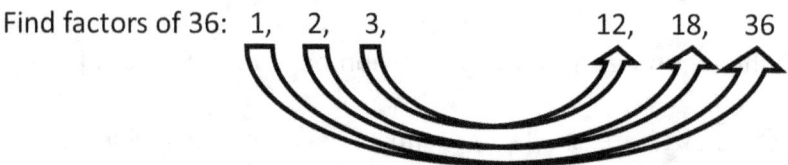

Then try dividing 36 by 4 which gives 9. List the pair 4 and 9 like this:

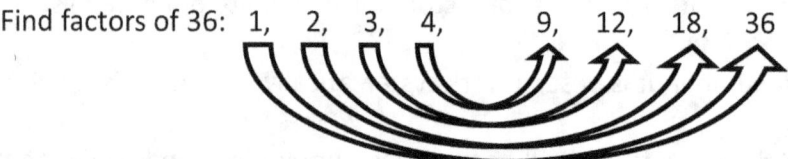

Then try dividing 36 by 5, but 5 doesn't divide evenly into 36. It leaves a remainder of 1. So, 5 is not a factor. We try the next number dividing 36 by 6 which gives 6. In this case the factor pairs up with itself.

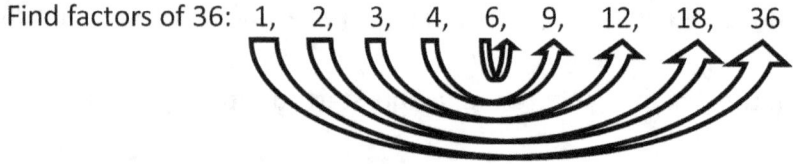

We have reached the "middle" of the list. We don't need to check any further for factors because it would have been paired up already with a number in the first half of the list. In fact, you only need to check factors up to the square root of the original number. So, for example for 36, you only need to check up to $\sqrt{36} = 6$. If we wanted to check for factors of 50, we would check up to $\sqrt{50} = 7.07$, so we could stop after checking 7. The factors of 36 are 1, 2, 3, 4, 6, 9, 12, 18, 36.

Practice Factors

List all positive factors for each number. The first one is done for you.

1) 10   1, 2, 5, 10

2) 8

3) 6

4) 9

5) 18

6) 12

7) 15

8) 16

9) 20

10) 14

11) 28

12) 22

13) 30

14) 25

15) 24

16) 26

17) 46

18) 55

19) 62

20) 54

21) 50

22) 60

23) 928

24) 845

25) 795

26) 835

27) 565

28) 976

29) 526

30) 735

Multiples of Numbers in Arithmetic

A multiple of a number is a number that is obtained by multiplying that number by any whole number. For example, the multiples of 3 are 3, 6, 9, 12, 15, and so on, because each of these numbers can be obtained by multiplying 3 by a whole number. Multiples can also be zero or negative. Multiples can be thought of as counting by that number, and the list of multiples is infinite because we can keep counting as high as we want.

The key difference between factors and multiples is that factors are the numbers that divide a given number without leaving a remainder, whereas multiples are the numbers obtained by multiplying a given number by any whole number. Another important distinction is that the factors of a number are always smaller than or equal to the number itself, while the multiples of a number can be larger than the number.

Practice Multiples

List the first 10 multiples of each number

1) Multiples of 2: Count by 2's all even numbers
   2, 4, 6, 8, 10, 12, 14, 16, 18, 20

2) Multiples of 3:

3) Multiples of 4:

4) Multiples of 5:

5) Multiples of 6:

6) Multiples of 7:

7) Multiples of 8:

8) Multiples of 9:

9) Multiples of 10:

10) Multiples of 11:

Prime Numbers

A prime number is a number that has exactly two unique factors. The number 2 is a prime number because 2 has exactly two factors: 1 and 2. A composite number is a number that has more than two factors such as 4 which is divisible by 1, 2, and 4. The number 1 is not considered to be a prime because 1 has only one factor, namely itself.

The Sieve of Eratosthenes is an ancient Greek algorithm (or process) for finding all prime numbers up to any given number. First, make a table of all numbers from 1 to the given number, let's say 100. The number 1 is not a prime, so we cross it out. The number 2 is the first prime, so we can circle it, but then we cross out all multiples of 2. Counting by 2's all other even numbers are crossed out. Then 3 is not crossed out, so it is our next prime, we can circle it and cross out any remaining multiples of 3. The number 4 is crossed out from being a multiple of 2. So, 5 is the next prime, cross out all other multiples of 5 and continue the process until you reach the end of the table.

Here's the table started after 2 and 3. Can you finish the table?

1	2	3	4	5	6	7	8	9	10
11	12	13	14	15	16	17	18	19	20
21	22	23	24	25	26	27	28	29	30
31	32	33	34	35	36	37	38	39	40
41	42	43	44	45	46	47	48	49	50
51	52	53	54	55	56	57	58	59	60
61	62	63	64	65	66	67	68	69	70
71	72	73	74	75	76	77	78	79	80
81	82	83	84	85	86	87	88	89	90
91	92	93	94	95	96	97	98	99	100

Here's what the Sieve of Eratosthenes looks like up to the number 100:

1	2	3	4	5	6	7	8	9	10
11	12	13	14	15	16	17	18	19	20
21	22	23	24	25	26	27	28	29	30
31	32	33	34	35	36	37	38	39	40
41	42	43	44	45	46	47	48	49	50
51	52	53	54	55	56	57	58	59	60
61	62	63	64	65	66	67	68	69	70
71	72	73	74	75	76	77	78	79	80
81	82	83	84	85	86	87	88	89	90
91	92	93	94	95	96	97	98	99	100

Creating tables in this way is quite tedious, so why would anyone want to do this? Prime numbers are so important in many areas of mathematics including arithmetic, number theory, algebra, cryptography, and more. Mathematicians keep looking for better ways of determining if a number is prime or composite.

Mathematicians consider primes to be the basic building blocks of all numbers. The Fundamental Theorem of Arithmetic, also known as the Unique Factorization Theorem or Prime Factorization Theorem, states that every positive integer can be represented uniquely as a product of prime numbers. So just like in chemistry every substance can be broken up into molecules and those molecules are built of individual atoms, primes are the atoms of numbers and essential to understanding algebra at a deeper level.

The prime factorization can be found using a factor tree method or division method.

Example 1: Find the prime factorization of each for the number 36.

Solution:

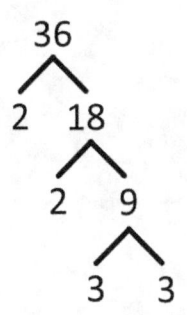

Factor Tree

Division Method

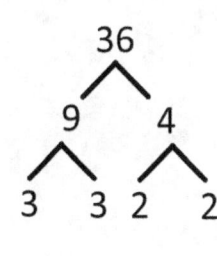

Even if you start off a different way, you end up with the same prime factors

So we can see that the prime factorization of 36, if we put the factors in order, is:

$$36 = 2 \times 2 \times 3 \times 3 = 2^2 \cdot 3^2$$

Practice Prime Factorization

Write the prime-power factorization of each. The first one is done for you.

1) 24 $24 = \boxed{2^3 \cdot 3}$ 2) 18

3) 14   4) 21

5) 16   6) 12

7) 15   8) 25

9) 20

10) 22

11) 90

12) 55

13) 75

14) 85

15) 82

16) 84

17) 58

18) 80

19) 65

20) 60

21) 813

22) 760

23) 656

24) 753

25) 717

26) 556

27) 579

28) 836

29) 581

30) 750

Areas of Rectangles

Experimenting and discovery are important ways that we learn mathematics. We can use the ideas we have learned so far about multiplication, factors, and prime numbers to solve new types of problems. Geometry is another area of mathematics where the tools of factoring will become important and help make the ideas more concrete and practical. In this section we will be using what we've learned so far to solve problems involving area.

Area is the amount of space inside a shape measured in square units such as square inches, sq. feet, cm^2, m^2, etc. If you were remodeling a room, you would measure the perimeter to determine how much molding would go around the room, while you would use the area to determine how much paint you would need. Here we need more formulas:

Rectangle: $A = lw$

Square: $A = s^2$

Here's an example: If we have a rectangle that is 2 ft by 4 ft, then it would contain 8 square feet.

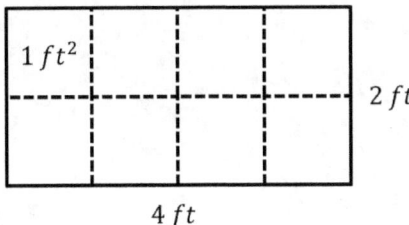

Practice Area of Rectangles

Find the area of each. The first one is done for you.

1)

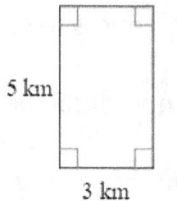

$5 \times 3 = 15 \text{ km}^2$

5 km

3 km

2)

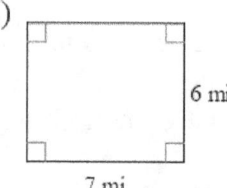

6 mi

7 mi

3)

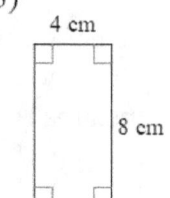

4 cm

8 cm

4)

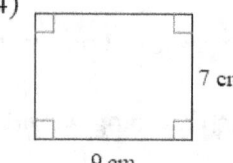

7 cm

9 cm

5)

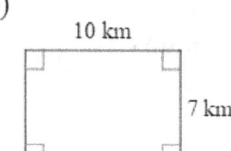

10 km

7 km

6)

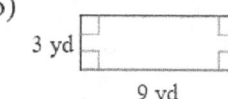

3 yd

9 yd

7)

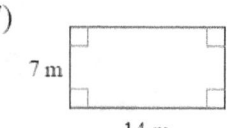

7 m

14 m

8)

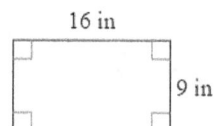

16 in

9 in

9)

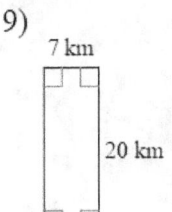

7 km

20 km

10)

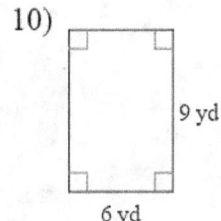

9 yd

6 yd

11)

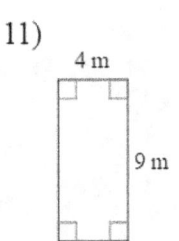

4 m

9 m

12)

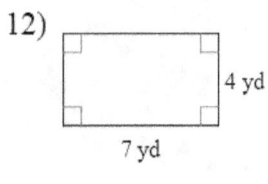

4 yd

7 yd

Find the missing measurement. The first one is done for you.

13)

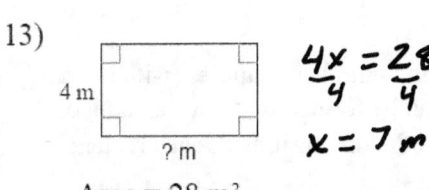

4 m

? m

$$\frac{4x}{4} = \frac{28}{4}$$

$$x = 7 m$$

Area = 28 m²

14)

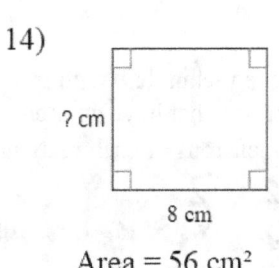

? cm

8 cm

Area = 56 cm²

15)

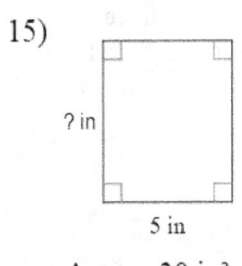

? in

5 in

Area = 30 in²

16)

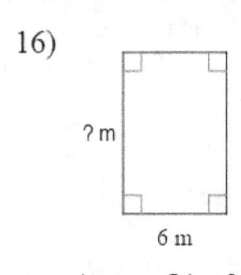

? m

6 m

Area = 54 m²

17)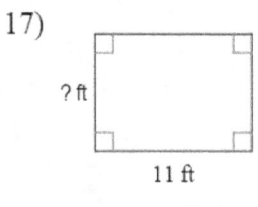

? ft

11 ft

Area = 88 ft²

18)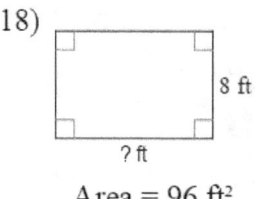

8 ft

? ft

Area = 96 ft²

19)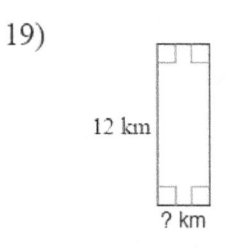

12 km

? km

Area = 48 km²

20)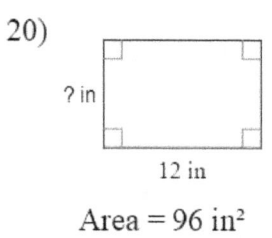

? in

12 in

Area = 96 in²

21) How many unique rectangles (with integer sides) can be formed that have an area of 12? Find the dimensions of each rectangle.

1×12
2×6   } 3 rectangles
3×4
~~4×3~~ repeat

22) How many unique rectangles (with integer sides) can be formed that have an area of 14? Find the dimensions of each rectangle.

23) How many unique rectangles (with integer sides) can be formed that have an area of 17? Find the dimensions of each rectangle.

24) How many unique rectangles (with integer sides) can be formed that have an area of 20? Find the dimensions of each rectangle.

25) How many unique rectangles (with integer sides) can be formed that have an area of 24? Find the dimensions of each rectangle.

26) How many unique rectangles (with integer sides) can be formed that have an area of 25? Find the dimensions of each rectangle.

27) How many unique rectangles (with integer sides) can be formed that have an area of 30? Find the dimensions of each rectangle.

28) How many unique rectangles (with integer sides) can be formed that have an area of 31? Find the dimensions of each rectangle.

29) How many unique rectangles (with integer sides) can be formed that have an area of 48? Find the dimensions of each rectangle.

30) How many unique rectangles (with integer sides) can be formed that have an area of 49? Find the dimensions of each rectangle.

31) What can you say about the areas where only 1 rectangle can be formed using integer side lengths?

32) What can you say about the areas where only 2 rectangle can be formed using integer side lengths?

Least Common Multiple (LCM)

We can find the Least Common Multiple LCM of two numbers such as 2 and 3. Multiples of 2 are 2, 4, 6, 8, 10, 12, 14, 16, 18, 20, 22, 24 ... Multiples of 3 are 3, 6, 9, 12, 15, 18, 21, 24 ... Some of the numbers are the same on both lists, namely 6, 12, 18, 24, and there would be an infinite of matches if we kept writing both lists. The least common multiple (LCM) is the smallest match which for 2 and 3 would be 6.

Practice Least Common Multiple

Find the LCM of each. The first one is done for you.

1) 12, 20
   12: 12, 24, 36, 48, 60
   20: 20, 40, 60, 80   60

2) 9, 6

3) 8, 16

4) 6, 8

5) 8, 12

6) 10, 6

7) 18, 12

8) 6, 14

9) 12, 16

10) 15, 10

11) 15, 20

12) 14, 12

Greatest Common Factor (GCF)

We can find the Greatest Common Factor GCF of two numbers such as 24 and 60. The factors of 24 are 1, 2, 3, 4, 6, 8, 12, 24. The factors of 60 are 1, 2, 3, 4, 5, 6, 10, 12, 15, 20, 30, 60. The number of factors is limited. Again, there are several numbers in both lists: 1, 2, 3, 4, 6, and 12. The greatest common factor (GCF) is the largest match which in this case would be 12.

Practice Greatest Common Factor

Find the GCF of each. The first one is done for you.

1) 18, 6   18: 1, 2, 3 6, 9, 18   2) 16, 20
   6: 1, 2   3, 6   GCF = 6

3) 6, 20   4) 15, 9

5) 11, 3   6) 16, 8

7) 15, 18   8) 5, 14

9) 12, 15   10) 18, 12

11) 15, 20   12) 20, 14

Finding the LCM and GCF Using a Venn Diagram

Usually, when we find LCM and GCF, we make lists as we did above. However, there is a way to find both, which is especially easier for larger numbers. We will factor both numbers completely and place the numbers in a Venn Diagram. The product of all the numbers in the intersection (the overlap of the two circles) will be the GCF. While the product of all the numbers overall, the union, will be the LCM.

In this example we find the LCM and GCF of 48 and 60. We make a factor tree for each. They have 2, 2, and 3 in common. 48 has two extra factors of 2, while 60 has an extra factor of 5. The GCF is 12 and the LCM is 240. When filling in the Venn Diagram start filling in the intersection with the factors that are shared with both numbers. Then fill in the remaining factors.

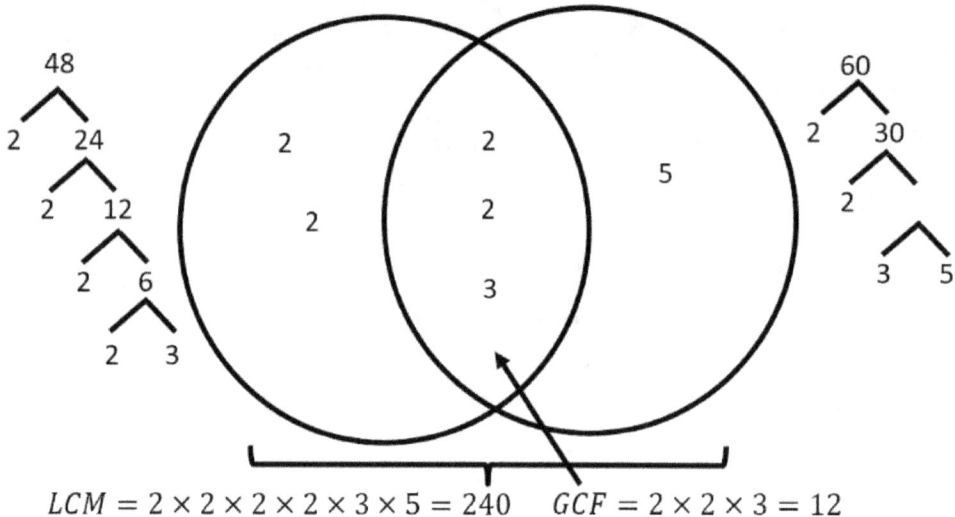

$$LCM = 2 \times 2 \times 2 \times 2 \times 3 \times 5 = 240 \quad GCF = 2 \times 2 \times 3 = 12$$

Here's a Venn Diagram for 40 and 80. Notice that if no other number can go in a section of the Venn Diagram, you can put the number 1 there. Here we see that 40 is a factor of 80.

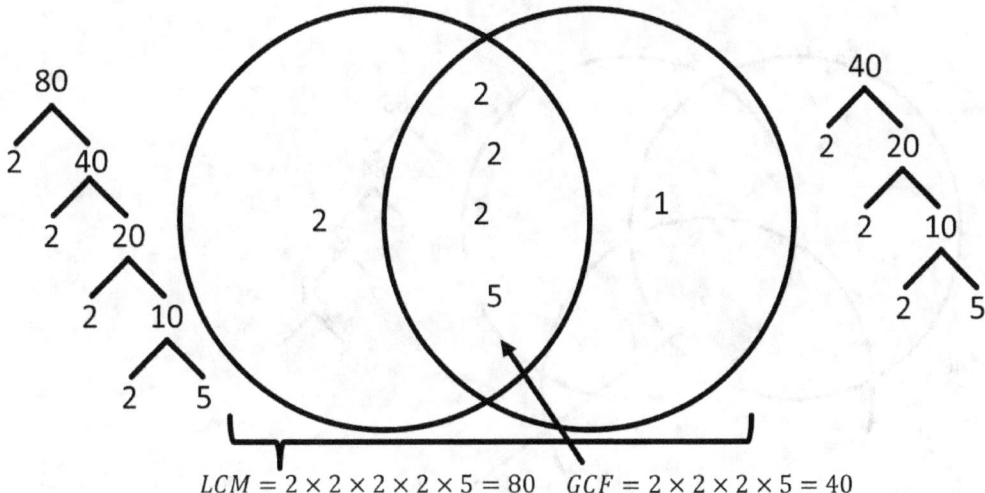

$$LCM = 2 \times 2 \times 2 \times 2 \times 5 = 80 \quad GCF = 2 \times 2 \times 2 \times 5 = 40$$

Here's a Venn Diagram for 50 and 63. Here 50 and 63 do not share any factors. We say that

they are relatively prime or **coprime**, therefore, the GCF is 1.

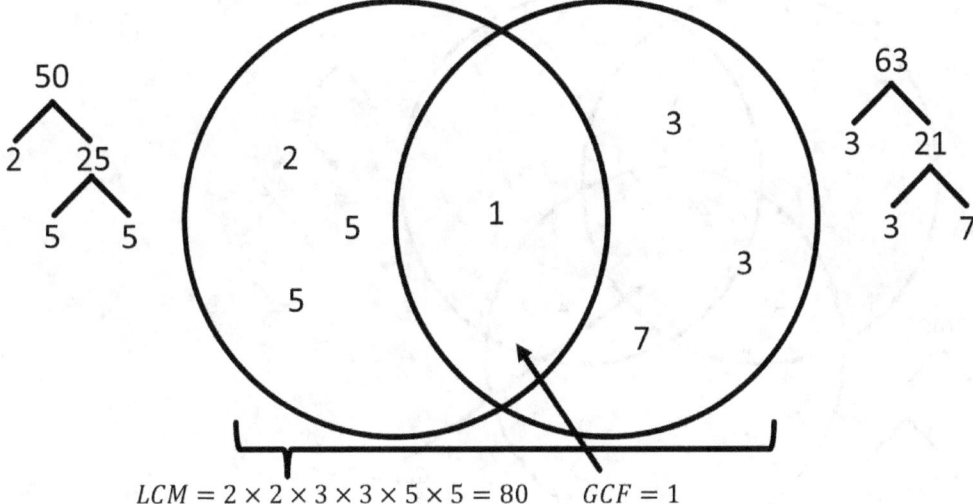

$$LCM = 2 \times 2 \times 3 \times 3 \times 5 \times 5 = 80 \quad GCF = 1$$

The factor tree method can also be used to find the LCM and GCF of three numbers. Here's the

process for finding the LCM and GCF of 140, 168 and 210.

Step 1: Make a factor tree for each number and determine what is shared among all three. Put

those numbers in the intersection of all three circles. As you do this it will help to cross them off

of the lists to keep track. The product of these numbers is the GCF.

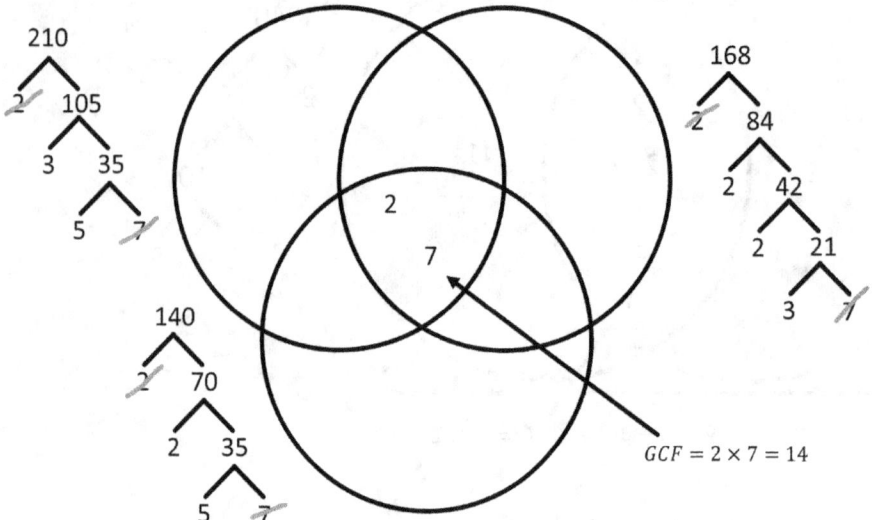

Step2: Now put in the factors that are shared between just two of the original numbers.

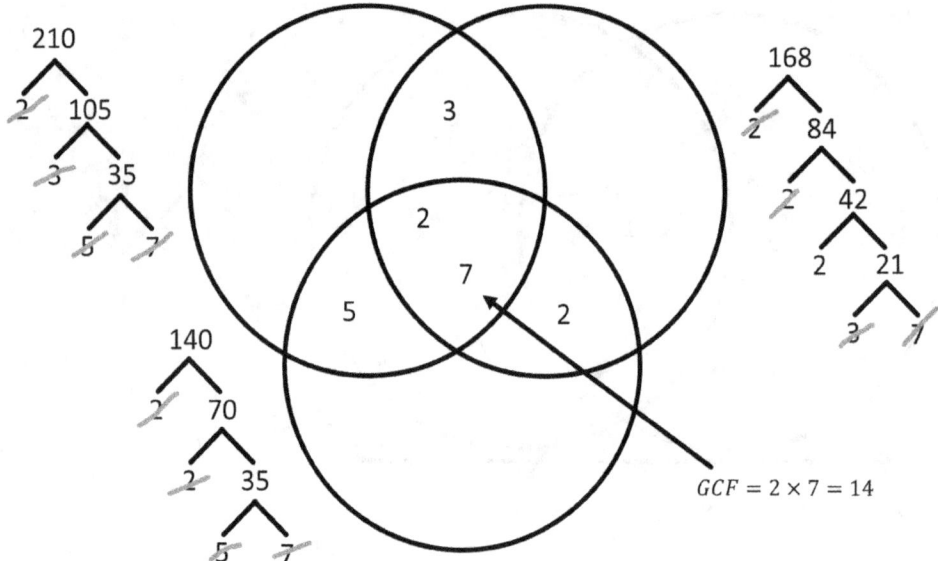

Step 3: Lastly, place any numbers that are not shared and are only factors of one number. If there is no number to place, then put a 1 in that section. Multiply all the numbers in the union of all the circles to get the LCM.

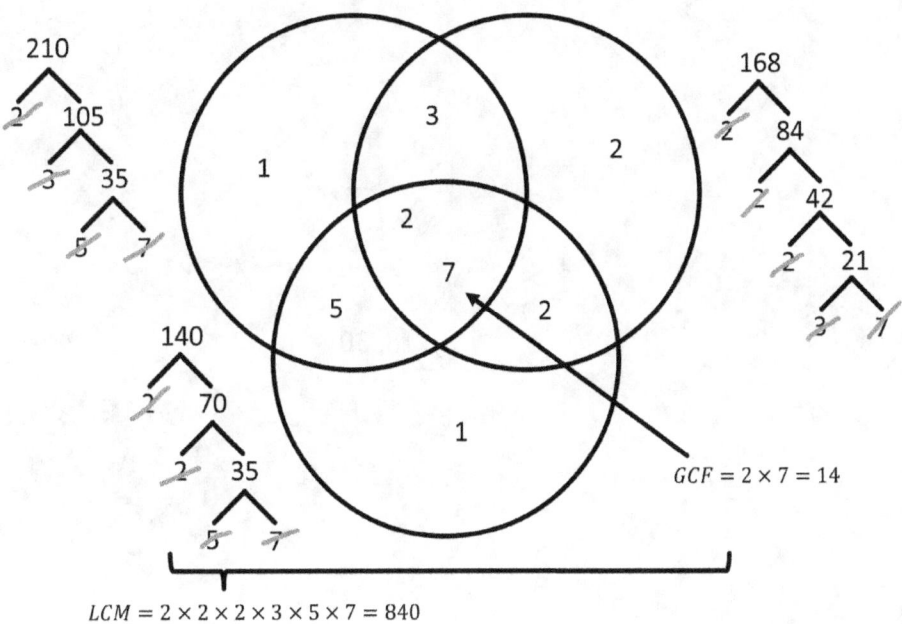

$$LCM = 2 \times 2 \times 2 \times 3 \times 5 \times 7 = 840$$

Practice LCM and GCF Using a Venn Diagram

Find the LCM and GCF of each using a Venn Diagram. The first one is done for you.

1) 16, 40

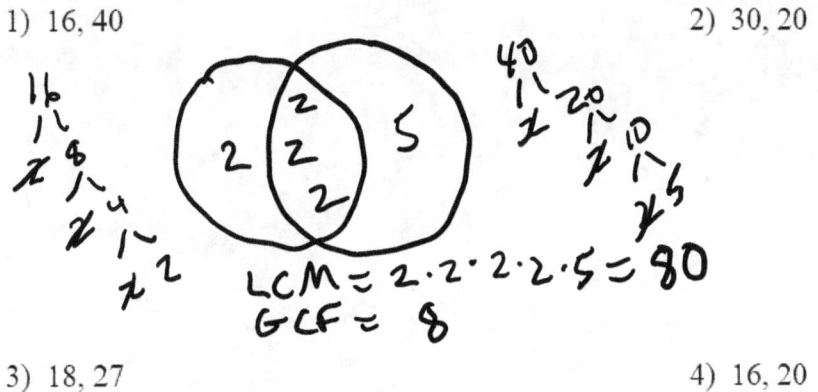

2) 30, 20

3) 18, 27

4) 16, 20

5) 40, 32

6) 10, 15

7) 16, 32

8) 18, 30

9) 36, 27

10) 40, 30

11) 42, 28

12) 60, 48

13) 22, 33

14) 30, 35

15) 25, 55

16) 56, 35

17) 56, 16

18) 42, 56

19) 55, 30

20) 22, 55

21) 60, 36, 24

22) 8, 36, 12

23) 15, 45, 35

24) 55, 44, 33

25) 55, 22, 33

26) 40, 25, 30

27) 56, 32, 24

28) 24, 36, 12

29) 45, 36, 27

30) 60, 24, 44

31) 72, 27, 18

32) 96, 72, 48

Algebraic Expressions

So far, we have been focused on the foundational arithmetic ideas that we will need to conquer more difficult algebraic concepts. Algebra is a mathematical branch that uses symbols to represent and manipulate unknown quantities to provide solutions to more complex problems than basic arithmetic alone can solve. A **variable** is a letter that represents a number, like x, y, n, etc. A **constant** is a known number, like $2, -1, \pi$, etc. A **coefficient** is a number that is being multiplied by a variable and written in front of the variable, like the 2 in $2x$. An algebraic **expression** is a combination of variables, numbers, and mathematical operations.

A **term** is part of an algebraic expression separated by a plus sign. For example, $2x^2 + 3y + 7$ has three terms: $2x^2$, $3y$, 7. Subtraction can be rewritten as addition, adding the opposite. So, we can write $5x^2 - 2xy + 3y - 7$ as $5x^2 + (-2xy) + 3y + (-7)$. We can see it has 4 terms: $5x^2$, $-2xy$, $3y$, -7. **Like terms** have the same variables and exponents. $2x^2$ and $5x^2$ are like terms. $3x^2$ and $7x$ are *not* like terms. We can simplify an expression by combining (adding/subtracting) like terms. When we combine like terms we add (or subtract) the coefficients (numbers in front) and keep the variable part the same. The variable part is like a label that doesn't change when we combine like terms.

(2a + p + 3b) + (2a + 2p + 2b) = 4a + 3p + 5b

For example, in $5x - 2y + 3 + 8x + 6y - 9$ we can combine like terms. Underlining like terms:

$$5x - 2y + 3 + 8x + 6y - 9 = 13x + 4y - 6$$

Algebraic expressions can be represented visually by rectangles and squares called algebra tiles. We can represent each unit with a small square and each variable x with a rectangle. Negative units will be shaded squares and negative x will be a shaded rectangle. A positive and negative of the same side will cancel each other out.

x	1	$-x$	-1

A visual representation and physical model such as these can make the concepts in algebra more concrete and clearer.

For instance, represent and simplify the following expression using the algebra tiles:

$3x - 4 - 2x + 6$

Practice Combining Like Terms

Simplify each expression by combining like terms. The first one is done for you.

1) $\underline{1} + \underline{3m} + \underline{1m} - \underline{4}$ 4m − 3

2) $a - 8 + 1 - 9a$

3) $1 + 3n + 7 - 10n$

4) $2n + 4 + n + 7$

5) $9k + 6 + k - 4$

6) $7x - 10 + x + 2$

7) $1 - 10b + b + 6$

8) $8x + 9 + 8x - 3$

9) $1 - 4n + 4n + 9$

10) $3 + 10m + 1 - 9m$

11) $-8 + 7x + 10 + 4x$

12) $9 + 2x + 4 - 5x$

13) $4 + x + 5 - 3x$

14) $5 - 9b + 10b + 9$

15) $2m^2 + 7m + 7 - 3m^2 + 7 - 6m$

16) $n^2 + n - 9 - n^2 + 5n + 1$

17) $x^2 + 4x + 1 + 6 + 7x + x^2$

18) $3x^2 + 1 + 8x + 1 + 2x^2 - 6x$

19) $-6x - 4x^2 - 10 + 5 - 3x + 2x^2$

20) $-7k + 5x^2 + 10 + 10 - 5k + x^2$

Properties of Exponents

To start working with algebraic expressions, we need to understand what the exponents for variables mean and what the rules are for manipulating exponents under different operations. Just as multiplication is repeated addition, exponentiation is repeated multiplication. For example, $3^4 = 3 \cdot 3 \cdot 3 \cdot 3$ or $x^5 = x \cdot x \cdot x \cdot x \cdot x$

$$a^m = \underbrace{a \cdot a \cdot a \cdots a}_{m \text{ times}}$$

What happens when we start raising exponents to exponents or doing other operations with exponents? It helps to start by breaking down what the exponents mean.

For example, $x^2 \cdot x^3 = x \cdot x \cdot x^3 = x \cdot x \cdot x \cdot x \cdot x = x^5$, this gives us a sense of what happens when we raise a multiply two monomials that have the same base. It appears that the exponents will be added.

On the other hand, take: $(x^2)^3 = x^2 \cdot x^2 \cdot x^2 = x \cdot x \cdot x \cdot x \cdot x \cdot x = x^6$, now we know what happens when we raise a power to a power. It appears that the exponents will be multiplied.

There are several other properties that are used in algebra, below is a table outlining the major properties of exponents.

Properties of Exponents				
Property Name	Notation	Explanation	Numeric Example	Algebraic Example
Product of Powers	$a^m \cdot a^n = a^{m+n}$	When multiplying with the same base, add the exponents	$2^3 \cdot 2^2 = 2^5$	$x \cdot x^5 = x^6$
Power of a Power	$(a^m)^n = a^{mn}$	When raising a power to a power, multiply the exponents	$(3^2)^3 = 3^6$	$(y^3)^4 = y^{12}$
Power of a Product	$(ab)^m = a^m b^m$	When a product is raised to a power, the power will go to each factor inside.	$(2 \cdot 3)^4 = 2^4 3^4$	$(2x^2 y^3)^3 = 8x^6 y^9$
Negative Exponents	$a^{-m} = \dfrac{1}{a^m}$	Negative exponents means to take reciprocal, flip & positive	$5^{-2} = \dfrac{1}{5^2} = \dfrac{1}{25}$	$x^{-1} = \dfrac{1}{x}$
Zero Exponent	$a^0 = 1, a \neq 0$	Anything* to the zero is one *Anything but zero	$2023^0 = 1$	$x^0 = 1$
Quotient of Powers	$\dfrac{a^m}{a^n} = a^{m-n}, a \neq 0$	When dividing with the same base, subtract the exponents	$\dfrac{2^7}{2^4} = 2^{7-4} = 2^3 = 8$	$\dfrac{x^3}{x} = x^2$
Power of a Quotient	$\left(\dfrac{a}{b}\right)^m = \dfrac{a^m}{b^m}, b \neq 0$	When a quotient is raised to a power, the power will go to each part inside.	$\left(\dfrac{2}{3}\right)^4 = \dfrac{2^4}{3^4}$	$\left(\dfrac{5x}{3y^3}\right)^2 = \dfrac{25x^2}{9y^6}$

Practice Properties of Exponents

Simplify. Your answer should contain only positive exponents.

1) $8 \cdot 8^4$ $\;\;= 8^1 \cdot 8^4 = 8^5$

2) $8^2 \cdot 8^3 \cdot 8^2$

3) $\dfrac{(-6)^2}{-6}$

4) $\dfrac{4^2}{4^4}$

5) $\left(2^3\right)^4$

6) $\left(5^2\right)^4$

7) $\left(\dfrac{(-3)^3}{(-3)^{-4} \cdot (-3)^3} \right)^{-1}$

8) $\dfrac{4}{\left(4^4\right)^2 \cdot 4^2}$

9) $7xy^2 \cdot x^2 y^3$

10) $5x^2 y^4 \cdot 4x^2 y^3$

11) $8a^3 b^4 \cdot 2ba^3$

12) $7x^4 y^2 \cdot x^4 y^2 \cdot 7xy^3$

13) $\dfrac{5yx^3}{6y^4}$

14) $\dfrac{x^2 y^2}{8x^3}$

15) $\dfrac{6x^2 y^4}{5xy^4}$

16) $\dfrac{5u^3 v^2}{7uv^4}$

17) $\left(6x^3 y^4\right)^3$

18) $\left(5n^4\right)^3$

19) $\left(5m^3 n^2\right)^2$

20) $\left(4v^4\right)^3$

21) $3x^3y^3 \cdot 5xy^2$

22) $2u^{-1} \cdot 2u^3v^4$

23) $7x \cdot 8x$

24) $6x^{-4}y^{-1} \cdot 3y^0$

25) $\dfrac{8nm^{-3}}{2m^3n^4}$

26) $\dfrac{2ab^3}{7a^{-1}b^{-4}}$

27) $\dfrac{4x}{5x^{-3}y^{-2}}$

28) $\dfrac{4x^2y^{-2}}{7y^4}$

29) $\left(3b^3\right)^{-3}$

30) $\left(2u^4v^3\right)^3$

31) $\left(5ab^{-1}\right)^{-2}$

32) $\left(2x^4y^{-4}\right)^3$

33) $\left(\dfrac{4m^0n^3}{3m^4n^4 \cdot 4m^3n^3}\right)^4$

34) $\dfrac{\left(yx^{-3}\right)^4}{2y^4 \cdot yx^3}$

35) $\left(\dfrac{4u^{-4} \cdot 3u^{-1}v^0}{u^{-3} \cdot 3u^{-2}v^2}\right)^0$

36) $\dfrac{\left(4x^0\right)^0}{yx^{-2} \cdot 4x^0y^2}$

37) $\left(\dfrac{3u^3v^3 \cdot 4u^0v^2}{\left(uv^3\right)^3}\right)^0$

38) $\dfrac{x^{-4}y^4}{\left(3y\right)^3 \cdot 3y}$

39) $\dfrac{4a^4b^{-1} \cdot 3ab^{-1}}{\left(3a^4b^{-1}\right)^4}$

40) $\dfrac{\left(3b^3\right)^4 \cdot \left(4a^3b^{-2}\right)^{-1}}{2a^2b^4}$

Factoring Monomials

A monomial is an algebraic expression that is one term ("mono-" means one) which is a combination of a coefficient (a number) and one or more variables raised to an exponent or power.

For example, $15x^2y^3$ is a monomial which is 15 multiplied by some unknown number x raised to the second power, then multiplied by another unknown number y raised to the third power. This monomial can be broken down into its prime factors. $15x^2y^3 = 3 \cdot 5 \cdot x \cdot x \cdot y \cdot y \cdot y$. Here x and y are considered to be primes since we have no additional information to factor them.

Practice Factoring Monomials

Write the prime factorization of each. Do not use exponents. The first one is done for you.

1) $8r^2s^3$ $= 2 \cdot 2 \cdot 2 \cdot r \cdot r \cdot s \cdot s \cdot s$

2) $22a^4b$

3) $16x^2y^3$

4) $10r^2s^5$

5) $30x^2yz^3$

6) $15mp^3q^2$

7) $24x^2y^4z^3$

8) $9x^2y^4z^3$

9) $12m^2n^2p^3$

10) $26x^4y^2z$

Factoring GCF from Monomials

When factoring out a GCF from monomials, first we look at the coefficients and take the GCF for all coefficients. Then look at the variable factors that are common to all the monomials, always taking the lowest power of each variable.

For example, say we want to find the GCF of $12x^2y^3z$ and $30x^4y$. Looking at the coefficients 12 and 30 the GCF is 6. Then the x's we have x^2 and x^4, so we write down the lowest being x^2. Then the y's we have y^3 and y, so we write down the lowest being y. Finally, the z's, but here the second monomial doesn't contain any z's, so the GCF will not contain any either. Therefore, GCF ($12x^2y^3z$, $30x^4y$) = $6x^2y$.

Practice Factoring GCF from Monomials

Find the GCF of each. The first one is done for you.

1) $38a^2, 44a^2$ **$2a^2$**

2) $44x^2, 44xy$

3) $45xy, 18y$

4) $16x, 16x^3$

5) $24x^2y, 24xy^2$

6) $18x^3, 30x^3$

7) $30x^2, 42x^4$

8) $10a, 2b$

9) $21yx^2, 49y^2x^2$

10) $21ab, 36a^2b$

11) $45x^2y^2, 27y^2$

12) $30n^2, 36mn^2$

9) $21yx^2, 49y^2x^2$

10) $21ab, 36a^2b$

11) $45x^2y^2, 27y^2$

12) $30n^2, 36mn^2$

13) $24y^2x^2, 50x^3y$

14) $12x^4y^3, 22x^2y^2$

15) $33uv^2w, 24uvw^2$

16) $44v^2uw^2, 44uw^3$

17) $48u^3vw^2$, $24u^3vw^4$

18) $42m^2n^2p^4$, $24m^2np^5$

19) $42x^3y^4z$, $49x^6y^4$

20) $26u^2vw^3$, $8v^2w^4$

Distributive Property

The distributive property is one of the most important properties algebra. The distributive property will allow us to simplify more types of expressions and lead us into factoring. Remember that the distributive property works with multiplication and addition. If we have 2(3+5) instead of adding in the parentheses first, we can distribute the 2 to the inside. This gives us 2(3+5) = 2(3) + 2(5) = 6 + 10 = 16. When we are dealing with just numbers it might seem crazy to solve a simple math problem this way, but distributive property starts to shine when we start introducing variables into the mix.

For example, 4(2x + 5) = 4(2x) + 4(5) = 8x + 20.

Negative numbers can also be distributed. Remember to distribute the negative sign!

$-3(2x - 4) = -3(2x) - 3(-4) = -6x + 12$

Although rarely seen, the distributive property also works if the parentheses are on the right and the multiplication is on the right, as seen here:

(3x + 4)5 = (3x)5 + 4(5) = 15x + 20.

Here's another using more terms and higher exponents:

$3x(2x^2 - 4x + 7) = 6x^3 - 12x^2 + 21x$

Distributive Property

We can also use distributive property and combining like terms in the same problem:

$$2(3x - 4) - 3(4x + 5) = 2(3x) - 2(4) - 3(4x) - 3(5)$$

Practice Distributive Property

Simplify each expression. The first one is done for you.

1) $-8(5 - 6x)$ $-40 + 48x$

2) $7(8b - 7)$

3) $4(3r + 8)$

4) $-(1 + 2n)$

5) $2(10 + 2x)$

6) $2(-10 + 3x)$

7) $6n(7n + 8)$

8) $5m^3(7m - 7)$

9) $3x^4(4x + 2)$

10) $2p^4(7p + 7)$

11) $5(5x - 1)$

12) $6x^2(3x + 4)$

13) $2(4k^2 - 4k + 5)$

14) $5(5p^2 - 2p - 7)$

15) $8(2m^2 - 2m - 1)$

16) $7(3k^2 + 5k + 6)$

17) $2x^3(x^2 + 4x - 8)$

18) $3m^2(5m^2 - 3m - 4)$

19) $6p(6p^2 + 4p - 2)$

20) $3n^3(2n^2 - 5n + 6)$

21) $3n^5(n^2 - 4n - 7)$

22) $5a^2(8a^2 - 7a - 7)$

23) $6(9n - 7) - 9(n - 8)$

24) $-6(n + 9) - 8(-6 + 7n)$

25) $-7(9 - 10a) + 4(a + 8)$

26) $-7(n - 1) - 10(6n + 5)$

27) $-6(10x - 2) + 10(4x - 2)$

28) $-3(n + 4) - 7(7n - 2)$

29) $-10(1 - 5n) + 9(1 - 3n)$

30) $8(x - 4) - 7(1 - 10x)$

31) $10(1 + n) - 7(10n + 3)$

32) $-(n - 5) + 10(2 - 3n)$

33) $3a(3a + 4) - 5(3a + 4)$

34) $5p(4p - 3) + 7(4p - 3)$

35) $4n(7n - 3) - 6(7n - 3)$

36) $7b(5b - 3) - 3(5b - 3)$

Distributive Property with Algebra Tiles

We can represent the distributive property by setting up a rectangle with dimensions equal to the factors that we want to multiply. The number and type of algebra tiles that make up the rectangle will give us the simplified expression. The number in front of the parentheses determines the number of rows and the terms inside the parentheses will tell us about the columns.

Let's take: $4(2x + 3)$

Here's what the algebra tile representation would look like:

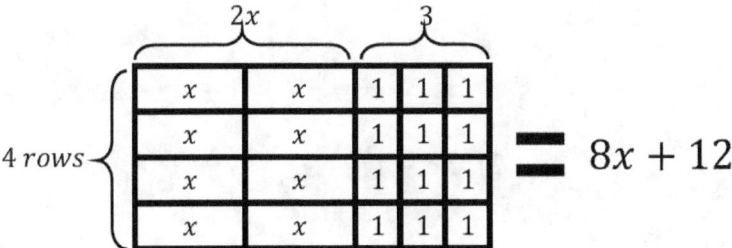

$$8x + 12$$

This also works with negatives. Remember that negative tiles will be shaded.

Here's what the algebra tile representation would look like for $2(-3x - 4)$:

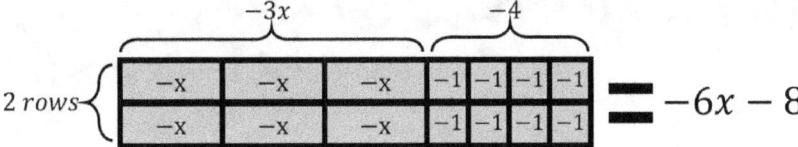

$$-6x - 8$$

We can also distribute with a negative number on the outside. It might seem like it doesn't

make much sense to have "negative rows," but we just have to remember the negative sign

when we multiply. A negative multiplied by a positive is negative and a negative multiplied by a

negative is a positive. Try $-4(3x - 2)$:

$$-12x + 8$$

Practice Distributive Property with Algebra Tiles

Simplify each expression.

1) $2(x+2)$

$=2x+4$

2) $2(x+1)$

3) $2(x-3)$

4) $2(-x+3)$

5) $2(3x-2)$

6) $3(x+2)$

7) $2(x-2)$

8) $3(x-2)$

9) $3(3x+2)$

10) $3(2x-1)$

11) $2(3x+3)$

12) $2(1+2x)$

Factoring Out a Greatest Common Factor

Factoring is just using the distributive property backwards. Instead of multiplying

$a(b + c)$ and getting a product of $ab + ac$, we work backwards from a product of the form

$ab + ac$ and factor out a GCF of a to get the product $a(b + c)$.

Let's factor $35x^3 - 56x^2 + 14x$. The GCF of the coefficients 35, -56 and 14 is 7. Now we

look at the variables x³, x², x, the lowest variable with the lowest exponent is x. So, the GCF is

7x. Each term needs to be divided by the GCF. We can now factor as:

$$35x^3 - 56x^2 + 14x = 7x(5x^2 - 8x + 2)$$

We are just at the beginning of factoring, but as we progress throughout all the different

factoring methods and shortcuts, it all comes down to this: The first step to factoring is to look

for a GCF. Let me say it again in case you missed it: **THE FIRST STEP TO FACTORING IS TO LOOK**

FOR A GCF.

Practice Factor Out GCF

Factor the common factor out of each expression. The first one is done for you.

1) $-21n^2 + 35n$ $7n(-3n + 5)$

$$\text{or } -7n(3n - 5)$$

2) $-80 + 24v$

3) $7x + 14$

4) $10p^3 - 6p^2$

5) $56n^2 - 24n$

6) $15r^2 + 21r$

7) $36n^3 - 18n - 27$

8) $63m^2 + 36m + 90$

9) $48n^3 + 80n^4 + 8n^5$

10) $40b^2 - 16b^3 - 48b^4$

11) $6n^2 - 9n + 2$

12) $30n^4 + 30n^3 - 20$

13) $30x^7y^2 + 25x^9 + 40x^7y$

14) $7x^5y^3 + 2x^5y + 3x^3$

15) $30y^6 - 30y^4x + 70y^3$

16) $20x^2y^8 + 24x^2y^6 + 20xy^6$

17) $8x^2y + 8x^3y - 2x^5y$

18) $20x^3y^4 + 20x^2y^2 - 12x^2y$

19) $10x^4z^3y^5 - 40x^2z^5y^2 + 45x^4z^2 + 5x^2zy^2$

20) $21k^4 + 35k^5h^3 + 21kh^5j - 14kj^3$

21) $-60x^3y^4z + 30x^6z - 54x^3z^2 + 12x^3y$

22) $90ab^5 - 80ab - 10a^5b - 50ac^3$

23) $12c^2b^4 - 12c^2 + 60c^4a + 42c^3a^2$

24) $10pr^5 + 20p^2r^3 - 35pr^4 + 45pr^2q^2$

25) $18n^4 + 108n^3 - 81n^2 + 63n$

26) $-70m^3np + 110$

27) $24x^9z^2y^4 + 12x^5zy^3 + 30x^2zy + 36x^2z$

28) $8a^3b^6 + 72a^3b^3 - 16a^2b^4$

29) $6 - 10hk + 11j^2k + 14j^2$

30) $12xy^3z^3 + 60xz^2 - 30xy$

31) $-735q^4r^8p^2 + 882q^2r^9 - 980qr^8p^2 + 98q^2r^8$

32) $-289y^8x^3z + 323y^7z^3 + 391y^4x + 272y^4$

Multiplying Binomials Leading Coefficient of 1

A binomial is an algebraic expression that has two terms ("bi-" meaning two). That is a binomial will have 2 parts separated by an addition sign, or if it is separated by a subtraction sign, we can rewrite it as adding the opposite. $2x + 3$, $4x^2 - 3y$, $-6x + 2y$ are three examples of binomials. Most binomials that will be encountered in algebra will be linear binomials with one variable of the form: $Ax + B$, where A and B are integers. In this section we focus exclusively on binomials with a leading coefficient of 1. When multiplying two binomials we take the first term from the first binomial and distribute it to each term in the second binomial. Then we distribute the second term from the first binomial to each term in the second binomial. Sometimes this procedure is called the **FOIL** method, standing for **F**irst, **O**utside, **I**nside, **L**ast.

First

$(x + 4)(x - 2) = x^2$

Outside

$(x + 4)(x - 2) = x^2 - 2x$

Inside

$(x + 4)(x - 2) = x^2 - 2x + 4x$

Last

$(x + 4)(x - 2) = x^2 - 2x + 4x - 8$

After multiplying each of the terms together, the two middle terms can be combined.

$$(x + 4)(x - 2) = x^2 - 2x + 4x - 8 = x^2 + 2x - 8$$

Here's another example:

$$(4x - 3)(2x + 5) = 8x^2 + 20x - 6x - 15 = 8x^2 + 14x - 15$$

Practice Multiplying Binomials Leading Coefficient of 1

Find each product.

1) $(x - 4)(x + 7)$ $x^2 + 7x - 4x - 28$

$x^2 + 3x - 28$

2) $(x - 6)(x - 3)$

3) $(n + 2)(n + 8)$

4) $(n + 2)(n - 5)$

5) $(m + 3)(m - 4)$

6) $(b - 5)(b - 4)$

7) $(x + 11)(x + 5)$

8) $(n + 9)(n - 8)$

9) $(a - 7)(a + 2)$

10) $(n + 9)(n - 11)$

11) $(x + 2)(x + 7)$

12) $(m + 3)(m - 12)$

13) $(p-6)(p-11)$

14) $(a-4)(a-10)$

15) $(m-5)(m-11)$

16) $(b-10)(b-12)$

17) $(r+3)(r-9)$

18) $(m-12)(m+7)$

19) $(p+1)(p+9)$

20) $(r+3)(r+12)$

Factoring Trinomials of the Form x^2+bx+c

So, you remember the first step to factoring? The first step to factoring is to look for a GCF. Now we begin to look for patterns and shortcuts.

Factoring trinomials of the form $x^2 + bx + c$ will use the same process going backwards. We want to unFOIL the trinomial into a product of two binomials. To do this look closely at what happens when we multiply two binomials, so that we can see how we should work backwards.

Watch the signs! The signature (positive and negative signs) of the trinomial gives us information about the signs of the binomials. If all the signs are positive, then the binomials both have positive signs. If the last sign is positive but the middle sign is negative, then both

binomials are negative. If the last sign is negative, then the binomials have opposite signs.

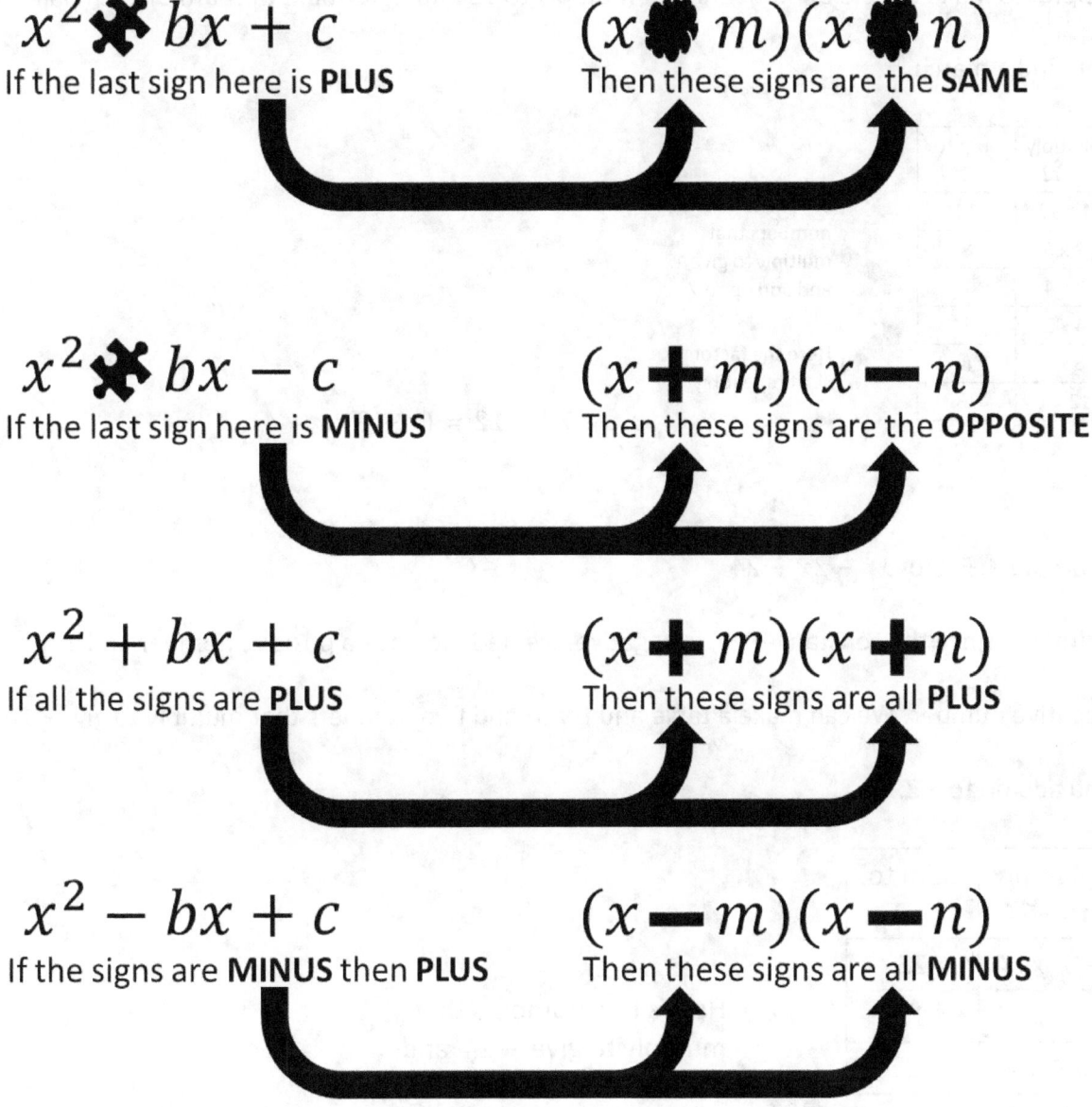

$$x^2 \text{ } bx + c$$
If the last sign here is **PLUS**

$$(x \text{ } m)(x \text{ } n)$$
Then these signs are the **SAME**

$$x^2 \text{ } bx - c$$
If the last sign here is **MINUS**

$$(x + m)(x - n)$$
Then these signs are the **OPPOSITE**

$$x^2 + bx + c$$
If all the signs are **PLUS**

$$(x + m)(x + n)$$
Then these signs are all **PLUS**

$$x^2 - bx + c$$
If the signs are **MINUS** then **PLUS**

$$(x - m)(x - n)$$
Then these signs are all **MINUS**

Example 1: Factor $x^2 + 7x + 12$

Solution: If this factors, it will factor into two binomials of the form $(x + m)(x + n)$ where m

and n are constants. If we foil, we get: $(x + m)(x + n) = x^2 + nx + mx + mn$. We can see that the product will equal the constant term and the sum is equal to the coefficient of x. We need to find two numbers whose product is 12 and whose sum is 7: *mn = 12 and m+n=7*. It is helpful to make a table of numbers that multiply to 12 and determine the sum of each pair until you find a match.

Multiply to 12	Sum to 7
1 x 12	13
2 x 6	8
3 x 4	7
4 x 3	7
6 x 2	8
12 x 1	13

We found two numbers that multiply to give 12 and add up to 7

Here the factors start repeating

$$x^2 + 7x + 12 = (x+3)(x+4)$$

Example 2: Factor $x^2 - 2x - 24$

Solution: Since the constant, -24, is negative, we need to have a positive number and a negative number. We can make a table and try to find two numbers that multiply to give -24 and add up to -2.

Multiply to − 24	Sum to − 2
− 1 x 24	23
− 2 x 12	10
− 3 x 8	5
− 4 x 6	2
− 6 x 4	− 2
− 8 x 3	− 5
− 12 x 2	− 10
− 24 x 1	− 23

Here's two numbers that multiply to give − 24 and add up to 2. We need − 2, so switch the signs

Here's the numbers that we want, multiplying to give − 24 and adding up to − 2.

$$x^2 - 2x - 24 = (x-6)(x+4)$$

Example 3: Factor $x^2 - 10x + 16$

Solution: What two numbers multiply to give 16 and add up to $-$ 10? How can two numbers multiply to give a negative and add to a positive? They must both be negative! If you can think of the numbers that work a table is not necessary. $x^2 - 10x + 16 = (x - 8)(x - 2)$

Example 4: Factor $x^2 - 10x + 12$

Solution:

If you can't think of the numbers, make a table to find them:

Multiply to 12	Sum to 6
1 x 12	13
2 x 6	8
3 x 4	7
4 x 3	7
6 x 2	8
12 x 1	13

We made a complete table, but there's no combination what will multiply to give 12 and add to 6. Therefore, we say that $x^2 - 10x + 12$ is unfactorable or prime.

Example 5: Factor $6x^3 + 12x^2 - 288x$ completely

Solution: The first step to factoring is to always look for a GCF. The GCF is *6x*. After taking out the GCF, we have $6x^3 + 12x^2 - 288x = 6x(x^2 + 2x - 48)$ Then continue factoring the trinomial, looking for factors of $-$ *48* that add up to *2*. After factoring the trinomial, you must still include the GCF on the outside as a factor. $(x^2 + 2x - 48) = (x + 8)(x - 6)$.

$6x^3 + 12x^2 - 288x = 6x(x^2 + 2x - 48) = 6x(x + 8)(x - 6)$.

Example 6: Find all the values of b such that the trinomial $x^2 + bx + 16$ can be factored.

Solution: Questions like these can easily be answered by making a table and looking at the numbers in the sum column.

Multiply to 16	Sum to ?
1 x 16	17
2 x 8	10
4 x 4	8
8 x 2	10

Here's where the table starts to repeat, so we can stop filling in the rest.

So, b = 8, 10, or 17

Practice Factoring Trinomials of the Form x^2+bx+c

Factor each completely.

1) $x^2 - 13x + 40$

M. to 40 | Add -13

-1 · -40	-41
-2 · -20	-22
-4 · -10	-14
-5 · -8	-13

$(x-5)(x-8)$

2) $n^2 - 3n - 70$

3) $k^2 - 10k + 16$

4) $v^2 + 9v + 20$

5) $n^2 - 3n - 40$

6) $n^2 + 11n + 10$

7) $r^2 - 5r - 50$

8) $x^2 - 8x + 15$

9) $v^2 - 10v + 24$

10) $x^2 + 8x - 20$

11) $6x^2 + 30xy - 300y^2$

12) $a^2 + ab - 56b^2$

13) $x^2 - 20xy + 100y^2$

14) $a^2 - 2ab - 3b^2$

15) $x^2 + 10xy + 16y^2$

16) $4x^2 - 324y^2$

17) $x^2 - 14xy + 48y^2$

18) $3a^2 - 3b^2$

19) $u^2 + 9uv + 18v^2$

20) $x^2 - 17xy + 70y^2$

21) $2x^2 + 12x - 80$

22) $3m^2 + 30m$

23) $5m^2 + 35m - 90$

24) $5x^2 - 40x + 60$

25) $3r^2 - 30r + 63$

26) $2a^2 - 8ab$

27) $2x^2 + 18xy - 20y^2$

28) $5a^2 + 5ab - 280b^2$

29) $2u^2 + 20uv + 48v^2$

30) $5u^2 - 30uv + 25v^2$

31) $2m^3 - 32m^2n + 96mn^2$

32) $5x^2y - 120xy^2 + 715y^3$

Squares and Cubes

There are several factoring patterns that use squares and cubes. It will be extremely helpful to memorize square numbers and cube numbers.

The square numbers from 1^2 to 20^2 are: *1, 4, 9, 16, 25, 36, 49, 64, 81, 100, 121, 144, 169, 196, 225, 256, 289, 324, 361, 400*. The cube numbers from 1^3 to 6^3 are: *1, 8, 27, 64, 125, 216*.

You can use square numbers and cube numbers to estimate decimal approximations for square roots of non-square numbers and cube roots of non-cube numbers.

Example 1: Estimate $\sqrt{12}$ to the nearest tenth

Solution: Find the closest squares that 12 is between: $\sqrt{9} < \sqrt{12} < \sqrt{16}$ or $3 < \sqrt{12} < 4$. 12 is closer to 9 than 16. So, a reasonable guess might be 3.4. The actual value is 3.46410161514...

Example 2: Estimate $\sqrt{350}$ to the nearest tenth

Solution: Find the closest squares that 350 is between: $\sqrt{324} < \sqrt{350} < \sqrt{361}$ or $18 < \sqrt{350} < 19$. 350 is 26 away from 324 and 11 away from 361. The total distance between the squares is 37. A reasonable guess might be 18.6 or 18.7. The actual value is 18.7082869338...

Practice Squares and Cubes

1) Estimate the value of $\sqrt{3}$ to the tenths place.

    1.7

2) Estimate the value of $\sqrt{8}$ to the tenths place.

3) Estimate the value of $\sqrt{5}$ to the tenths place.

4) Estimate the value of $\sqrt{14}$ to the tenths place.

5) Estimate the value of $\sqrt{7}$ to the tenths place.

6) Estimate the value of $\sqrt{6}$ to the tenths place.

7) Estimate the value of $\sqrt{10}$ to the tenths place.

8) Estimate the value of $\sqrt{2}$ to the tenths place.

9) Estimate the value of $\sqrt{22}$ to the tenths place.

10) Estimate the value of $\sqrt{24}$ to the tenths place.

11) Estimate the value of $\sqrt{19}$ to the tenths place.

12) Estimate the value of $\sqrt{12}$ to the tenths place.

13) Estimate the value of $\sqrt{26}$ to the tenths place.

14) Estimate the value of $\sqrt{17}$ to the tenths place.

15) Estimate the value of $\sqrt{31}$ to the tenths place.

16) Estimate the value of $\sqrt{47}$ to the tenths place.

17) Estimate the value of $\sqrt{34}$ to the tenths place.

18) Estimate the value of $\sqrt{28}$ to the tenths place.

19) Estimate the value of $\sqrt{60}$ to the tenths place.

20) Estimate the value of $\sqrt{66}$ to the tenths place.

21) Estimate the value of $\sqrt{57}$ to the tenths place.

22) Estimate the value of $\sqrt{52}$ to the tenths place.

23) Estimate the value of $\sqrt{90}$ to the tenths place.

24) Estimate the value of $\sqrt{39}$ to the tenths place.

25) Estimate the value of $\sqrt{75}$ to the tenths place.

26) Estimate the value of $\sqrt{105}$ to the tenths place.

27) Estimate the value of $\sqrt{43}$ to the tenths place.

28) Estimate the value of $\sqrt{130}$ to the tenths place.

29) Estimate the value of $\sqrt{190}$ to the tenths place.

30) Estimate the value of $\sqrt{140}$ to the tenths place.

31) Estimate the value of $\sqrt{175}$ to the tenths place.

32) Estimate the value of $\sqrt{150}$ to the tenths place.

33) Estimate the value of $\sqrt{260}$ to the tenths place.

34) Estimate the value of $\sqrt{160}$ to the tenths place.

35) Estimate the value of $\sqrt{220}$ to the tenths place.

36) Estimate the value of $\sqrt{540}$ to the tenths place.

37) Estimate the value of $\sqrt{420}$ to the tenths place.

38) Estimate the value of $\sqrt{300}$ to the tenths place.

39) Estimate the value of $\sqrt{380}$ to the tenths place.

40) Estimate the value of $\sqrt{118}$ to the tenths place.

Difference of Squares Pattern

An important factoring pattern is the difference of squares. Examine what happens when you multiply the two binomials and use the FOIL method we get:

$$(a + b)(a - b) = a^2 - ab + ab - b^2 = a^2 - b^2$$

The outside and inside terms cancel out and we are left with a difference of squares.

Example 1: Multiply $(x + 9)(x - 9)$

Solution: $(x + 9)(x - 9) = x^2 - 81$

Example 2: Multiply $(13 + x)(13 - x)$

Solution: $(13 + 9)(13 - 9) = 169 - x^2$

Example 3: Multiply $(5x + 4y)(5x - 4y)$

Solution: $(5x + 4y)(5x - 4y) = 25x^2 - 16y^2$

Example 4: Multiply $(3x^2 + 2y^3)(3x^2 - 2y^3)$

Solution: $(3x^2 + 2y^3)(3x^2 - 2y^3) = 9x^4 - 4y^6$

Practice Difference of Squares Pattern

Find each product.

1) $(n-6)(n+6)$ $n^2 - 36$

2) $(v+3)(v-3)$

3) $(x+2)(x-2)$

4) $(x+4)(x-4)$

5) $(n+8)(n-8)$

6) $(n+5)(n-5)$

7) $(k-4)(k+4)$

8) $(k-9)(k+9)$

9) $(8m-5)(8m+5)$

10) $(5x-7)(5x+7)$

11) $(3v+5)(3v-5)$

12) $(2x+3)(2x-3)$

13) $(4a - 1)(4a + 1)$

14) $(8n + 5)(8n - 5)$

15) $(4p - 6)(4p + 6)$

16) $(6b + 4)(6b - 4)$

17) $(5x + 8y)(5x - 8y)$

18) $(8a - 8b)(8a + 8b)$

19) $(6x - 2y)(6x + 2y)$

20) $(8y + 5x)(8y - 5x)$

21) $(5x + 5y)(5x - 5y)$

22) $(8x + 2y)(8x - 2y)$

23) $(7y + 6x)(7y - 6x)$

24) $(7x + 8y)(7x - 8y)$

25) $(14x + 6y)(14x - 6y)$

26) $(9x + 19y)(9x - 19y)$

27) $(18y - 17x)(18y + 17x)$

28) $(17u - 5v)(17u + 5v)$

29) $(15x + 18y)(15x - 18y)$

30) $(13x - 8y)(13x + 8y)$

31) $(3x - 13y)(3x + 13y)$

32) $(-2y + 15x)(-2y - 15x)$

Factoring Difference of Squares

Factoring a difference of squares involves writing an expression as the product of two binomials. $a^2 - b^2 = (a + b)(a - b)$. First, we identify the perfect squares a^2 and b^2, then take the square roots to get a and b. One binomial is a sum, and one is a difference.

Visual Proof:

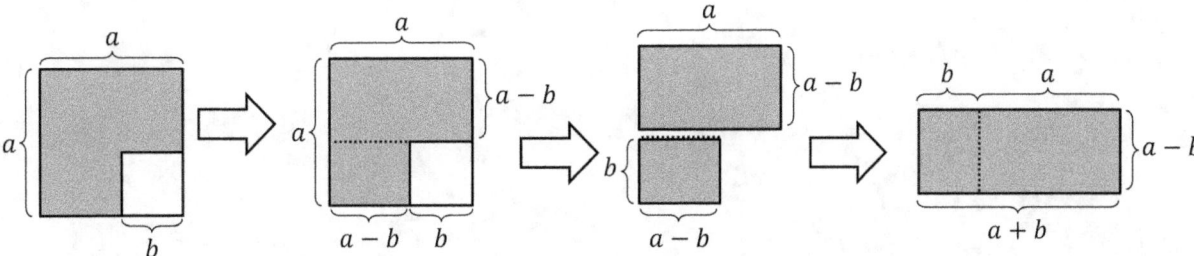

Factoring Difference of Squares

Example 1: Factor $x^2 - 49$

Solution: $x^2 - 49 = (x - 7)(x + 7)$

Example 2: Factor $25x^2 - 64y^2$

Solution: Here two variables are used, but we still have a difference of squares, and this can be

factored: $25x^2 - 64y^2 = (5x - 8y)(5x + 8y)$

Example 3: Factor $27x^3 - 75x$

Solution: In this example, it does not seem like we have a difference of squares but remember

the first rule to factoring: look for a GCF. After taking out a GCF, then continue factoring the

difference of squares: $27x^3 - 75x = 3x(9x^2 - 25) = 3x(3x - 5)(3x + 5)$

Example 4: Factor $36x^2 + 25$

Solution: Although $36x^2$ and 25 are both squares, this is a sum of squares not a difference of

squares. $36x^2 + 25$ is not factorable or prime.

Practice Factoring Difference of Squares

Factor each completely.

1) $b^2 - 100$ $\quad (b + 10)(b - 10)$

10^2

2) $x^2 - 49$

3) $v^2 - 25$

4) $n^2 - 36$

5) $x^2 + 25$

6) $b^2 - 81$

7) $x^2 - 64$

8) $n^2 - 4$

9) $100a^2 - 49b^2$

10) $64x^2 - 9y^2$

11) $16x^2 - 81y^2$

12) $4m^2 - n^2$

13) $64x^2 - 81y^2$

14) $49x^2 - 100y^2$

15) $49m^2 - 16n^2$

16) $4x^2 - 49y^2$

17) $36m^2 - n^2$

18) $32x^2 - 648y^2$

19) $25u^2 - 4v^2$

20) $40x^2 - 10y^2$

21) $16a^2 - 9b^2$

22) $100x^2 + 81y^2$

23) $150x^2 - 294y^2$

24) $x^2 - 81y^2$

25) $49x^2 - 121y^2$

26) $9x^2 - 289y^2$

27) $980x^2 - 845y^2$

28) $392m^2 - 242n^2$

29) $8n^2 - 72m^2$

30) $25x^2 - 144y^2$

31) $80x^4 - 4500y^4$

32) $1690m^4 + 640n^4$

Squaring Binomials

Now we will square binomials. When squaring a binomial, we multiply a binomial by itself. The distributive property or FOIL method can be used. After FOILing, the first and last terms will be squares and the middle terms get doubled up. The formula is written as

$(a \pm b)^2 = a^2 \pm 2ab + b^2$

Visual Proof:

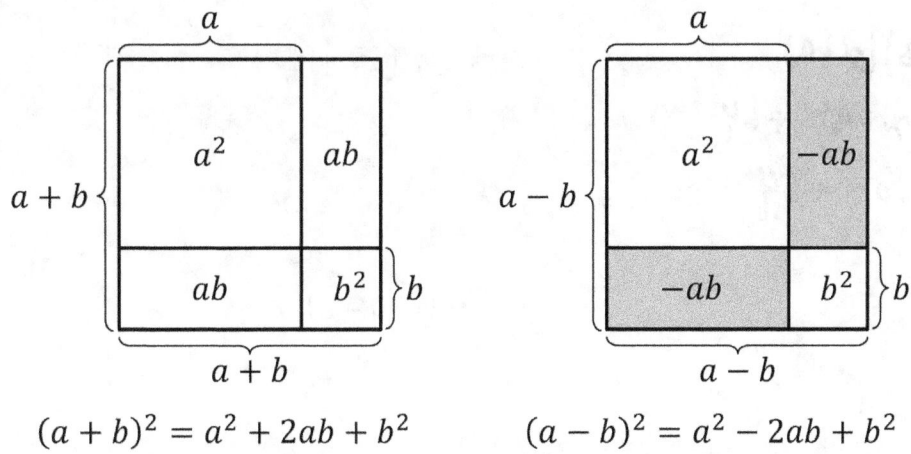

$$(a + b)^2 = a^2 + 2ab + b^2 \qquad (a - b)^2 = a^2 - 2ab + b^2$$

Example 1: Multiply $(a + 3)^2$

Solution: $(a + 3)^2 = (a + 3)(a + 3) = a^2 + 3a + 3a + 9$

Example 2: Multiply $(4b - 7)^2$

Solution: $(4b - 7)^2 = (4b - 7)(4b - 7) = 16b^2 - 28b - 28b + 49 = 16b^2 - 56b + 49$

Example 3: Multiply $(11x - 6y)^2$

Solution: $(11x - 6y)^2 = 121x^2 - 132xy + 36y^2$

Example 4: Multiply $(3x^4 + 2y^3)^2$

Solution: $(3x^4 + 2y^3)^2 = (3x^4 + 2y^3)(3x^4 + 2y^3)$

$= 9x^8 + 6x^4y^3 + 6x^4y^3 + 4y^6 = 9x^8 + 12x^4y^3 + 4y^6$

Practice Squaring Binomials

Find each product.

1) $(x + 8)^2 =$ $(x+8)(x+8)$
$$x^2 + 8x + 8x + 64$$
$$x^2 + 16x + 64$$

2) $(v + 1)^2$

3) $(v - 7)^2$

4) $(a - 8)^2$

5) $(b - 3)^2$

6) $(b + 7)^2$

7) $(p + 5)^2$

8) $(b - 6)^2$

9) $(2b + 5)^2$

10) $(8x - 5)^2$

11) $(11x - 3)^2$

12) $(5b - 7)^2$

13) $(9x - 11)^2$

14) $(6m + 7)^2$

15) $(2m - 1)^2$

16) $(11p - 7)^2$

17) $(10a + 3b)^2$

18) $(11x - 12y)^2$

19) $(5x - 8y)^2$

20) $(5u + 12v)^2$

21) $(5x - 7y)^2$

22) $(3y + x)^2$

23) $(4x + 11y)^2$

24) $(3x + 10y)^2$

25) $(2m - 15n)^2$

26) $(5v + 8u)^2$

27) $(5x + 16y)^2$

28) $(8y + 9x^2)^2$

29) $(12x - 7y^2)^2$

30) $(x^3 + 17y^2)^2$

31) $(12m + 13n^2)^2$

32) $(7m^2 - 20n^3)^2$

Factoring Perfect Square Trinomials

A perfect square trinomial is one that comes from squaring a binomial as we did in the last section. Now we factor in the opposite direction. When given a trinomial, if the first term is a square and the last term is a square, take the square roots of both. Then multiply those two numbers and double it. If the result matches with the middle term, then it is a perfect square. Note that the first and last terms should always be positive.

$a^2 \pm 2ab + b^2 = (a \pm b)^2$

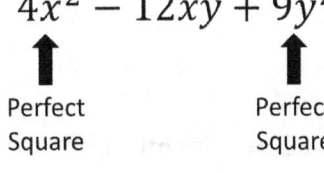

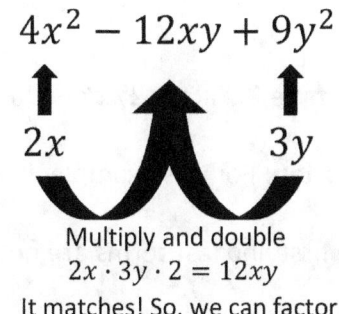

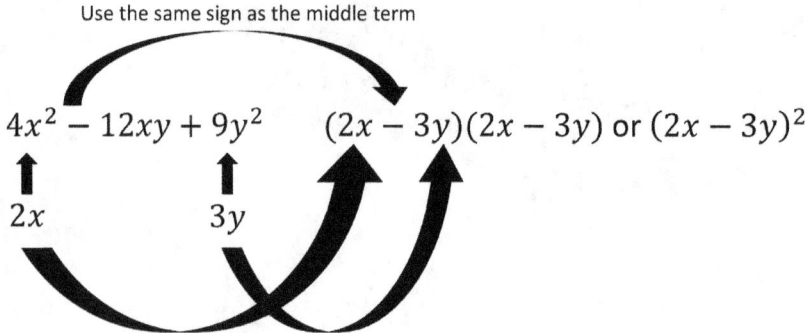

Here's an example of when the trinomial doesn't factor as a perfect square trinomial:

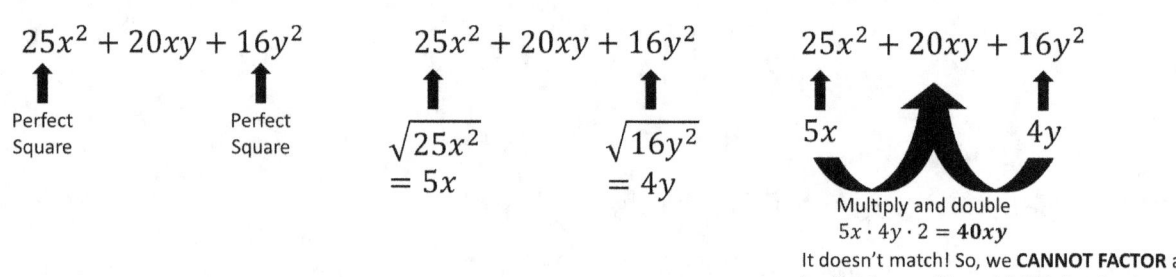

Example 1: Factor $25n^2 + 90n + 81$

Solution: Notice that the first term is a square and the last term is a square. Taking the square roots of both we get 5n and *9*. The product of these is *45n*. Doubling this becomes *90n*, which matches the middle term. Therefore, this follows the perfect square trinomial factoring pattern. To factor we use the square roots and the middle sign. $25n^2 + 90n + 81 = (5n + 9)^2$.

Example 2: Factor $27x^2 - 90xy + 75y^2$

Solution: For this example, it does not appear that we have a perfect square trinomial because the first and last terms are not squares. But recall the first step to factoring: **THE FIRST STEP TO FACTORING IS TO TAKE OUT A GCF**. Taking out the GCF is 3, we get:

then $27x^2 - 90xy + 75y^2 = 3(9x^2 - 30xy + 25y^2)$ Now we can see that the first term is a square and the last term is a square. Taking the square roots of both we get $3x$ and $5y$. The product of these is $15xy$. Doubling this becomes $30xy$, which matches the middle term (except for the negative sign, but that's ok). Therefore, use the perfect square trinomial factoring pattern. To factor we use the square roots and the middle sign:

$$27x^2 - 90xy + 75y^2 = 3(9x^2 - 30xy + 25y^2) = 3(3x - 5y)(3x - 5y) = 3(3x - 5)^2$$

Example 3: Factor $81a^2 + 36ab + 16b^2$

Solution: Again, the first term is a square and the last term is a square. Taking the square roots of both we get $9a$ and $4b$. The product of these is $36ab$. Doubling this becomes $72ab$, which does NOT match the middle term. Therefore, this trinomial does not follow the perfect square trinomial factoring pattern. There is no GCF and in general a polynomial that follows the "pattern" $a^2 \pm ab + b^2$ (where the middle term is not multiplied by 2) does not factor or is prime. $81a^2 + 36ab + 16b^2$ is not factorable or prime.

Example 4: Factor $49x^6 - 28x^3y^4 + 4y^8$

Notice that the first term is a square and the last term is a square. The terms that have variables are squares if the coefficient is a square and the variable's exponent is even. Taking the square roots of both we get $7x^3$ and $2y^4$. The product of these is $14x^3y^4$. Doubling this becomes $28x^3y^4$, which matches the middle term. Therefore, this follows the perfect square trinomial factoring pattern. To factor we use the square roots and the middle sign.

$$49x^6 - 28x^3y^4 + 4y^8 = (7x^3 - 2y^4)^2.$$

Practice Factoring Perfect Square Trinomials

Factor each completely.

1) $9x^2 - 60x + 100$ $(3x - 10)(3x - 10)$

$3x$ 10

$2 \cdot 3x \cdot 10$

2) $81x^2 + 18x + 1$

3) $49x^2 - 28x + 4$

4) $25n^2 + 20n + 4$

5) $25x^2 + 40x + 16$

6) $64n^2 + 112n + 49$

7) $49n^2 - 14n + 1$

8) $36v^2 - 60v + 25$

9) $100a^2 - 60ab + 9b^2$

10) $9x^2 - 60xy + 100y^2$

11) $49x^2 - 25y^2$

12) $9a^2 + 12ab + 4b^2$

13) $64u^2 - 16uv + v^2$

14) $196m^2 - 4n^2$

15) $112yx^2 - 392y^2x + 343y^3$

16) $49a^2 + 42ab + 9b^2$

17) $36a^2 + 132ab + 121b^2$

18) $81a^2 - 342ab + 361b^2$

19) $4536x^2 - 9576xy + 5054y^2$

20) $867yx^2 + 510y^2x + 75y^3$

21) $64m^2 + 272mn + 289n^2$

22) $1734b^2 - 2856ba + 1176a^2$

23) $49x^2 - 168xy + 144y^2$

24) $324m^2 - 468mn + 169n^2$

25) $y^4 + 2y^2x^2 + x^4$

26) $20y^4 + 100y^2x^2 + 125x^4$

27) $50nx^4 + 80nx^2y^2 + 32ny^4$

28) $27x^4 - 18x^2y^2 + 3y^4$

29) $27m^6 + 36m^3n^3 + 12n^6$

30) $9u^6 - 12u^3v^2 + 4v^4$

31) $16x^4 - 40x^2y^3 + 25y^6$

32) $12y^2x^6 - 36y^5x^3 + 27y^8$

Multiplying Polynomials with More than Two Terms

When multiplying a monomial by a polynomial we simply distribute the monomial to all the terms inside. When multiplying a binomial by a binomial, we also distribute, but this is sometimes called FOILing, where FOIL is an acronym for **F**irst, **O**utside, **I**nside and **L**ast. When multiplying a binomial and a trinomial, we distribute every term in the binomial by every term in the trinomial. It may help to line up the polynomials vertically as if you multiplied two large numbers.

Example 1: Multiply $(2m + 3)(2m^2 + 2m - 7)$

Solution: We can set this up vertically to make it easier to combine the like terms:

$$\begin{array}{r} 2m^2 + 2m - 7 \\ \times \quad\quad 2m + 3 \\ \hline -21 \end{array}$$

$$\begin{array}{r} 2m^2 + 2m - 7 \\ \times \quad\quad 2m + 3 \\ \hline 6m - 21 \end{array}$$

$$\begin{array}{r} 2m^2 + 2m - 7 \\ \times \quad\quad 2m + 3 \\ \hline 6m^2 + 6m - 21 \end{array}$$

$$2m^2 + 2m - 7$$
$$\times \qquad 2m + 3$$
$$\overline{}$$
$$6m^2 + 6m - 21$$
$$-14m$$

$$2m^2 + 2m - 7$$
$$\times \qquad 2m + 3$$
$$\overline{}$$
$$6m^2 + 6m - 21$$
$$4m^2 - 14m$$

$$2m^2 + 2m - 7$$
$$\times \qquad 2m + 3$$
$$\overline{}$$
$$6m^2 + 6m - 21$$
$$+ \quad 4m^3 + 4m^2 - 14m$$
$$\overline{}$$
$$4m^3 + 10m^2 - 8m - 21$$

$$(2m + 3)(2m^2 + 2m - 7) = 4m^3 + 10m^2 - 8m - 21$$

Example 2: Multiply $(2u^2 + 3uv - 6v^2)(4u^2 + 5uv - 6v^2)$

Solution:

$$2u^2 + 3uv - 6v^2$$
$$\times\ 4u^2 + 5uv - 6v^2$$
$$\overline{}$$
$$-12u^2v^2 - 18uv^3 + 36v^4$$

$$2u^2 + 3uv - 6v^2$$
$$\times\ 4u^2 + 5uv - 6v^2$$
$$\overline{}$$
$$-12u^2v^2 - 18uv^3 + 36v^4$$
$$10u^3v + 15u^2v^2 - 30uv^3$$

$$2u^2 + 3uv - 6v^2$$
$$\times\ 4u^2 + 5uv - 6v^2$$
$$\overline{}$$
$$-12u^2v^2 - 18uv^3 + 36v^4$$
$$10u^3v + 15u^2v^2 - 30uv^3$$
$$8u^4 + 12u^3v - 24u^2v^2$$
$$\overline{}$$
$$8u^4 + 22u^3v - 21u^2v^2 - 48uv^3 + 36v^4$$

$$(2u^2 + 3uv - 6v^2)(4u^2 + 5uv - 6v^2) = 8u^4 + 22u^3v - 21u^2v^2 - 48uv^3 + 36v^4$$

Practice Multiplying Polynomials More than Two Terms

Find each product.

1) $(7r + 3)(5r^2 + 5r + 3)$

$$35r^3 + 35r^2 + 21r$$
$$15r^2 + 15r + 9$$
$$\overline{}$$
$$35r^3 + 50r^2 + 36r + 9$$

2) $(2x - 6)(3x^2 - 3x + 5)$

3) $(7k - 2)(2k^2 + 6k + 2)$

4) $(5n + 8)(7n^2 - 5n - 4)$

5) $(u + 7v)(7u^2 - 4uv - v^2)$

6) $(8u + 8v)(7u^2 + 2uv + 2v^2)$

7) $(7x + 6y)(5x^2 - 5xy + 7y^2)$

8) $(x - 7y)(4x^2 - 5xy + 2y^2)$

9) $(5b^2 + 4b - 4)(b^2 - 6b + 5)$

10) $(5x^2 - 4x - 8)(4x^2 + 8x - 1)$

11) $(2k^2 + k + 7)(7k^2 + 8k - 5)$

12) $(4a^2 + 7a - 1)(a^2 - 3a - 4)$

13) $(11a^2 - 5ab + 9b^2)^2$

14) $(2x^2 + 7xy - 7y^2)(2x^2 + 12xy - 6y^2)$

15) $(4x^2 - 6xy + 12y^2)(3x^2 + xy + 8y^2)$

16) $(6x^2 + 10xy + 6y^2)(2x^2 + 5xy - y^2)$

17) $(4x + 3)(16x^2 - 12x + 9)$

18) $(4m + 5)(16m^2 - 20m + 25)$

19) $(2u - 5)(4u^2 + 10u + 25)$

20) $(3u - 2)(9u^2 + 6u + 4)$

Factoring Sum/Difference of Cubes

With squares, we can factor a difference of squares but not a sum of squares. With cubes we can factor both a sum and difference of cubes. The pattern for both is similar and can be written in the formula: $a^3 \pm b^3 = (a \pm b)(a^2 \mp ab + b^2)$. Notice the signs switching. A mnemonic device to help you remember the signs is **SOAP** standing for **S**ame, **O**pposite, **A**lways **P**lus.

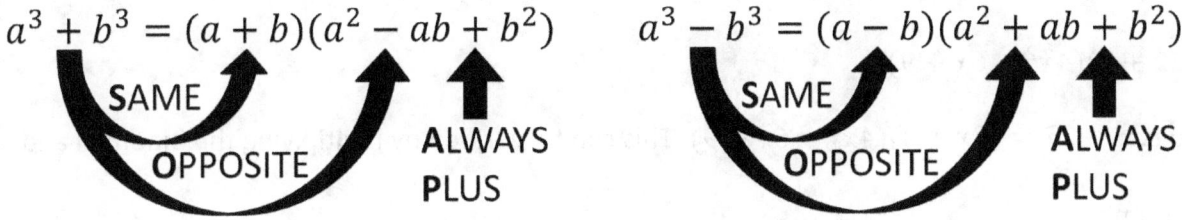

$$a^3 + b^3 = (a+b)(a^2 - ab + b^2) \qquad a^3 - b^3 = (a-b)(a^2 + ab + b^2)$$

SAME   OPPOSITE   ALWAYS PLUS   SAME   OPPOSITE   ALWAYS PLUS

You may wonder if the trinomial $a^2 \mp ab + b^2$ can be factored as it looks similar to a perfect square trinomial except the middle term is not doubled. In fact, if the degree (highest exponent) of the trinomial is 2, then $a^2 \mp ab + b^2$ will not be able to be factored. Whereas if it is 4 or higher it is possible that it might be factorable:

For example, $a^4 + a^2 + 1 = (a^2 - a + 1)(a^2 + a + 1)$.

Being able to recognize the cube numbers will help when factoring these: 1, 8, 27, 64, 125, 216, 343, 512, 729, 1000.

Example 1: Factor $8x^3 + 27$

Solution: Notice that the first term is a cube, and the last term is a cube. We will be able to factor any sum or difference of cubes. First set up the parentheses, then put in the signs using

SOAP. Since the sign is positive, the first sign will be the same, positive, the second sign is opposite, so it will be negative, the third sign will always be positive. The signs will look like this:

$(\ +\ )(\ -\ +\ )$

Taking the cube roots of both we get *2x* and *3*. These will be the terms of the binomial:

$(2x + 3)(\ -\ +\ )$

In the trinomial, the first and last terms will be the squares of 2x and 3, the middle term will be the product of 2x and 3:

$(2x + 3)(4x^2 - 6x + 9)$

So $8x^3 + 27 = (2x + 3)(4x^2 - 6x + 9)$. This can be verified by multiplying the binomial and trinomial.

Example 2: Factor $125x^3 - 64y^3$

Solution: $125x^3 - 64y^3 = (5x - 4y)(25x^2 + 20xy + 16y^2)$

Example 3: Factor $250x^4 + 432x$

Solution: First step to factoring is to always look for a GCF: $250x^4 + 432x = 2x(125x^3 + 216)$. Then continue to factor the sum of cubes, using SOAP. Make sure to include your GCF as part of your answer:

$250x^4 + 432x = 2x(125x^3 + 216) = 2x(5x + 6)(25x^2 - 30x + 36)$

Practice Factoring Sum/Difference of Cubes

Factor each completely.

1) $27x^3 + 125$

$(3x + 5)(9x^2 - 15x + 25)$

2) $8x^3 - 125$

3) $27 - 125x^3$

4) $125u^3 + 64$

5) $27u^3 - 8$

6) $125u^3 + 8$

7) $64x^3 + 125$

8) $x^3 - 64$

9) $729x^3 - 64y^3$

10) $125a^3 + 343b^3$

11) $-1000x^3 + 343y^3$

12) $8m^3 - 729n^3$

13) $729u^3 - 125v^3$

14) $216m^3 + 343n^3$

15) $729u^3 + 64v^3$

16) $216u^3 - 125v^3$

17) $x^3 - 1000y^3$

18) $yx^3 - 1000y^4$

19) $729yx^3 + 343y^4$

20) $1000x^3 - 343y^3$

21) $125m^3 - 216n^3$

22) $x^6y + 729x^3y^4$

23) $343a^3 + 1000b^3$

24) $-729y^2x^3 - 64y^5$

25) $375x^4 - 1536xy^3$

26) $27x^4 + 343xy^3$

27) $729x^3 + 8y^3$

28) $-750yx^3 - 1296y^4$

29) $343x^3 - 1000y^3$

30) $-8ba^3 - b^4$

31) $-343b^2a^4 - 27b^5a$

32) $1080yx^4 - 625y^4x$

Factoring GCF More Than One Term

We have factored out a GCF, and usually a GCF is just a monomial or one term. But our next factoring strategy will use factoring out a GCF that is a binomial. This GCF can be factored out and we write the remaining terms in the other factor.

$$5x(2x + 3) - 7(2x + 3) = (2x + 3)(5x - 7)$$

Example 1: Factor $3x(5x + 2) + 7(5x + 2)$

Solution: Notice that the first part has 5x +2 while the last part has 5x + 2. We can factor this out a GCF of 5x + 2 from both terms: $3x(5x + 2) + 7(5x + 2) = (5x + 2)(3x + 7)$.

Example 2: Factor $3x(2x - 5) - 2x + 5$

Solution: Notice that the first part has 2x – 5 while the last part has – 2x + 5. These are opposites of each other, and we can rewrite as $3x(2x - 5) - (2x - 5)$. Now factor out a GCF of 2x – 5: $3x(2x - 5) - (2x - 5) = (2x - 5)(3x - 1)$.

Factor by taking out a GCF with multiple terms.

1) $x(x+3)+6(x+3)$

$$(x+3)(x+6)$$

2) $x(x+2)+3(x+2)$

3) $x(x-5)-7(x-5)$

4) $x(x-9)-5(x-9)$

5) $3x(5x+7)+4(5x+7)$

6) $7x(4x-9)-3(4x-9)$

7) $3x(9x+5)-7(9x+5)$

8) $9x(4x-7)-2(4x-7)$

9) $2x(x-4)+x-4$

10) $x(x-7)+x-7$

11) $4x(x+2)+x+2$

12) $4x(3x-7)+9(3x-7)$

13) $5x(2x - 9) + 3(2x - 9)$

14) $9x(x + 7) + x + 7$

15) $6x(2x + 3) - 2x - 3$

16) $4x(3x + 2) - 3x - 2$

17) $4x(7x - 5) - 7x + 5$

18) $3x(5x - 2) - 5x + 2$

19) $3x(5x - 2) + 10x - 4$

20) $9x(4x - 7) + 8x - 14$

21) $5x(2x + 3) + 6x + 9$

22) $7x(4x + 1) + 16x + 4$

23) $x^2(5x - 2) + 3x(5x - 2) - 2(5x + 2)$

24) $2x^2(3x - 5) - 7x(3x - 5) + 4(3x - 5)$

25) $5x^2(2x + 7) - 4x(2x + 7) + 2x + 7$

26) $3x^2(7x - 3) + x(7x - 3) + 7x - 3$

27) $7x^2(4x - 5) + x(4x - 5) - 4x + 5$

28) $9x^2(5x - 3) + 2x(5x - 3) - 5x + 3$

29) $9x(x^2 + 2x + 3) - 4(x^2 + 2x + 3)$

30) $4x(x^2 - 3x + 5) - 7(x^2 - 3x + 5)$

31) $5x(x^2 - 7x + 9) + x^2 - 7x + 9$

32) $9x(x^2 - 4x + 7) - x^2 + 4x - 7$

Factoring by Grouping

Factoring by grouping is typically used when a polynomial has four terms. The first step to any factoring problem is always look for a GCF. Then group the terms into pairs. Ideally, you want to group the terms in such a way that each pair has a common factor. Usually, you can just group the first two terms together and the last two terms together, although sometimes you may need to switch the order. Factor out the common factors from each pair of terms. This is usually done by taking out the greatest common factor (GCF) of each pair. After factoring out the common factors from each pair, you should have two new expressions. Look for any common factors between the two new expressions. If there are any common factors, factor them out. Write the factored form of the original expression by combining the factored common factors from both groups.

Example 1: Factor $2x^3 + 4x^2 + 3x + 6$

Solution: There's no GCF, so we will put $2x^3$ and $4x^2$ together in the first group and $3x$ and 6 together in the second group:

$$2x^3 + 4x^2 + 3x + 6 = (2x^3 + 4x^2) + (3x + 6)$$

Now take out the GCF's for each group:

$$x^3 + 4x^2 + 3x + 6 = (2x^3 + 4x^2) + (3x + 6) = 2x^2(x + 2) + 3(x + 2)$$

Notice how (x +2) is in both parts, we can factor that out as a GCF and write the remaining factors in the second parentheses.

$$x^3 + 4x^2 + 3x + 6 = (2x^3 + 4x^2) + (3x + 6) = 2x^2(x + 2) + 3(x + 2) = (x + 2)(2x^2 + 3)$$

Example 2: Factor $96x^5 - 12x^4 + 144x^3 - 18x^2$

Solution: There's a GCF of $6x^2$,

$$96x^5 - 12x^4 + 144x^3 - 18x^2 = 6x^2(16x^3 - 2x^2 + 24x - 3)$$

$$= 6x^2((16x^3 - 2x^2) + (24x - 3))$$

Just keep the GCF outside, while working with the rest just ignore the GCF of $6x^2$. When you finish the problem just write it in front for your answer.

Take out the GCF's from each group:

$(16x^3 - 2x^2) + (24x - 3) = 2x^2(8x - 1) + 3(8x - 1) = (8x - 1)(2x^2 + 3)$. Now don't forget the GCF!

$$96x^5 - 12x^4 + 144x^3 - 18x^2 = 6x^2(8x - 1)(2x^2 + 3)$$

Example 3: Factor $21xy - 32y + 28x - 24y^2$

Solution: There's no GCF, so we will try to put 21x and −32y together in the first group and 28x and −24y² together in the second group:

$$21xy - 32y + 28x - 24y^2 = (21xy - 32y) + (28x - 24y^2)$$

Factoring by Grouping

Now try to take out the GCF's for each group:

$$(21xy - 32y) + (28x - 24y^2) = y(21x - 32) + 2(14x - 12y^2)$$

This does not seem to be working, the remaining factors do not match up. Try changing the order:

$$21xy + 28x - 32y - 24y^2 = (21xy + 28x) + (-32y - 24y^2) = 7x(3y + 4) - 8y(4 + 3y)$$

Now 3y + 4 and 4+3y are the same and can be factored out.

$$21xy + 28x - 32y - 24y^2 = 7x(3y + 4) - 8y(4 + 3y) = (3y + 4)(7x - 8y)$$

Example 4: Factor $81x^7y - 225x^6y^2 + 45x^7 - 405x^6y^3$

Solution: First look for a GCF which is 9x^6

$$81x^7y - 225x^6y^2 + 45x^7 - 405x^6y^3 = 9x^6(9xy - 25y^2 + 5x - 45y^3)$$

You can probably see that grouping the first two and last two will not work. So, let's rearrange:

$$(-25y^2 + 5x) + (9xy - 45y^3) = -5(5y^2 - x) + 9y(x - 5y^2)$$

Notice that the binomials in the parentheses are not the same, but they are opposites of each other. This means we need to switch the signs of one of the GCF's we took out.

$$(-25y^2 + 5x) + (9xy - 45y^3) = 5(-5y^2 + x) + 9y(x - 5y^2)$$

Now the binomials match up (just a different order), we can continue to factor.

$$81x^7y - 225x^6y^2 + 45x^7 - 405x^6y^3 = 9x^6\big((-25y^2 + 5x) + (9xy - 45y^3)\big)$$

$$= 9x^6\big(5(-5y^2 + x) + 9y(x - 5y^2)\big) = 9x^6(x - 5y^2)(5 + 9y)$$

Factor each completely.

1) $16k^3 - 20k^2 + 4k - 5$

$$4k^2(4k - 5) + 1(4k - 5)$$
$$(4k - 5)(4k^2 + 1)$$

2) $5x^3 + 35x^2 + 8x + 56$

3) $5a^3 + 40a^2 + 4a + 32$

4) $56x^3 + 7x^2 + 64x + 8$

5) $24a^3 - 20a^2 + 42a - 35$

6) $5x^3 - x^2 - 40x + 8$

7) $30v^3 - 12v^2 + 25v - 10$

8) $35n^3 - 21n^2 - 30n + 18$

9) $6ab + 8a + 21b + 28$

10) $10mu - 6mv - 35nu + 21nv$

11) $6xy - 4x - 3y^2 + 2y$

12) $56mn + 64m + 35n^2 + 40n$

13) $25uv + 10u + 5v^2 + 2v$

14) $15xy + 12x^2 - 20y - 16x$

15) $6mn + 30m + 7n + 35$

16) $24ab - 32a + 3nb - 4n$

17) $30pz - 36pc - 180qz + 216qc$

18) $32mn + 12m^2 - 56pn - 21pm$

19) $3mn + 3m + 2n + 2$

20) $40mn + 15m^4 + 16n + 6m^3$

21) $48xy - 6x + 8y - 1$

22) $14bz - 10bc + 35xz - 25xc$

23) $49xy - 21x - 7y^2 + 3y$

24) $96uv - 240um - 112mv + 280m^2$

25) $35ab + 40 + 25a + 56b$

26) $40xy + 6x - 48x^2 - 5y$

27) $8mu^2 + 35nv + 5mv + 56nu^2$

28) $30xy - 60x + 75x^2 - 24y$

29) $24m^2u^2 - 35n^2v - 28m^2v + 30n^2u^2$

30) $56ab - 8na - 64a^2 + 7nb$

31) $14mn + 40 - 8m - 70n$

32) $ab + 10 + 5a + 2b$

Factoring Trinomials of the Form ax^2+bx+c

Now we can factor the most difficult type of polynomials which are trinomials that don't start with a coefficient of 1. These trinomials will be of the form $ax^2 + bx + c$ and be factored into two binomials such that: $ax^2 + bx + c = (px + m)(qx + n)$

Then we have to find p, q, m and n such that: $a = pq$, $b = pn + qm$, and $c = mn$.

The task may seem a little daunting at first, but there are several ways of making it easier.

Factoring Trinomials of the Form ax^2+bx+c

Remember the first step to factoring is to look for a GCF. Then start from the "easy" side, the side with the fewest factors (primes are the best to start with). We can make educated guesses and check our results to find the binomial factors.

$ax^2 + bx + c$ $(px + m)(qx + n)$ $ax^2 - bx + c$ $(px - m)(qx - n)$ $ax^2 ✖ bx - c$ $(px + m)(qx - n)$
If all the signs are **PLUS**   Then these signs are all **PLUS**   If the signs are **MINUS** then **PLUS**   Then these signs are all **MINUS**   If the last sign here is **MINUS**   Then these signs are the **OPPOSITE**

Example 1: Factor $7x^2 + 19x - 6$

Solution: The first term is 7 and the last term is − 6. Which of these has fewer factors? Of course, 7, since it is a prime. So, start factoring from the first term. The signs are plus then minus, so the binomial will have opposite signs. We have two possible options:

$7x^2 + 19x - 6 = (7x + \quad)(x - \quad)\ or\ (7x - \quad)(x + \quad)$

Since the middle term is positive 19, it makes more sense if the 7x gets multiplied by a positive number, so the second option makes more sense. Now think about the factors of -6 and fill those in:

$(7x - 6)(x + 1)\ or\ (7x - 1)(x + 6)\ or\ (7x - 2)(x + 3)\ or\ (7x - 3)(x + 2)$

Which pair gives the middle term of 19x? $(7x - 2)(x + 3)$

Therefore, $7x^2 + 19x - 6 = (7x - 2)(x + 3)$

Example 2: Factor $24k^2 - 52k + 20$

Solution: The first step to factoring is to look for a GCF. Here the GCF is 4. Factor it out to get:

$24k^2 - 52k + 20 = 4(6k^2 - 13k + 5)$

Now look at the trinomial and determine if it can be factored again. Which side has fewer factors, 6 or 5? This time start by factoring the 5. The signs are negative then positive, so the signs must be the same and both be negative.

$(6k^2 - 13k + 5) = (\quad k - 5)(\quad k - 1)$

There are several ways to factor the 6:

$(6k - 5)(k - 1)\ or\ (k - 5)(6k - 1)\ \ or\ (2k - 5)(3k - 1)\ or\ (3k - 5)(2k - 1)$

Which set gives $-13k$ for the middle term?

$24k^2 - 52k + 20 = 4(6k^2 - 13k + 5) = 4(3k - 5)(2k - 1)$

This method works fairly well, especially if the coefficients are small or prime. But in general, it can be quite time consuming to guess and check through multiple possibilities.

A better approach is to factor by grouping. But that only works for polynomials with four terms and our polynomials have been trinomials with three terms. The secret is to split the middle term into two terms that make it possible to factor by grouping. To find out how to split the middle term we need to find factors of ac that add up to b.

Example 3: Factor $8x^2 - 25x + 18$

Solution: Notice there is no GCF. First make a table as we've done before. Find two numbers that multiply to give ac and that add up to b.

Factoring Trinomials of the Form ax^2+bx+c

Multiply to 8x18=144	Sum to −25
−1 x −144	−145
−2 x −72	−74
−3 x −48	−51
−4 x −36	−40
−6 x −24	−30
−8 x −18	−26
−9 x −16	−25
−12 x −12	−24

Here's where we find the two numbers that multiply to give ac and add up to b. Now we know how to split $-25x$, into $-9x$ and $-16x$

$$8x^2 - 25x + 18 = 8x^2 - 9x - 16x + 18$$

Factor by grouping

$$8x^2 - 25x + 18 = (8x^2 - 9x) + (-16x + 18) = x(8x - 9) - 2(8x - 9) = (8x - 9)(x - 2)$$

Check by FOILing

$$(8x - 9)(x - 2) = 8x^2 - 16x - 9x + 18 = 8x^2 - 25x + 18$$

Example 4: Factor $9x^2 - 9x - 70$

Solution:

Multiply to 9 x (− 70) = − 630	Sum to −9
1 x −630	−629
2 x −315	−313
3 x −210	−207
5 x −126	−121
6 x −105	−99
7 x −90	−83
9 x −70	−61
10 x −63	−53
14 x −45	−31
15 x −42	−27
18 x −35	−17
21 x −30	−9

Here's where we find the two numbers that multiply to give ac and add up to b. Now we know how to split $-9x$, into −$21x$ and $-30x$

$$9x^2 - 9x - 70 = 9x^2 - 30x + 21x - 70 = 3x(3x - 10) + 7(3x - 10)$$

$$= (3x - 10)(3x + 7)$$

Example 5: Factor $7x^2 + 28x + 20$

Solution: There's no GCF, so we multiply a and c to get 140. Try to find factors of 140 that add up to 28. Making a table works best to ensure we look at all possibilities:

Multiply to 7x 20 = 140	Sum to 28
1 x 140	141
2 x 70	72
4 x 35	39
5 x 28	33
7 x 20	27
10 x 14	24

If the factors of 140 could add up to 28 then it would be between these two entries. It's not on the list. Therefore, the polynomial is not factorable!

$7x^2 + 28x + 20$ is not factorable or prime!

Practice Factoring Trinomials of the Form ax^2+bx+c

Factor each completely.

1) $7b^2 + 57b - 54$

$7b^2 - 6b + 63b - 54$

$b(7b - 6) + 9(7b - 6)$

$(7b - 6)(b + 9)$

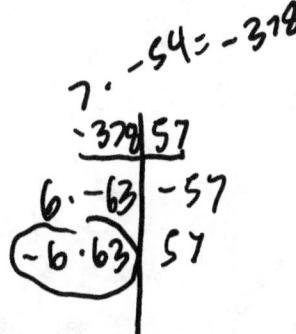

$7 \cdot -54 = -378$

$-378 \,|\, 57$

$6 \cdot -63 \,|\, -57$

$-6 \cdot 63 \,|\, 57$

2) $5n^2 - 7n - 24$

3) $2m^2 - 19m + 35$

4) $3k^2 + 14k + 8$

5) $10x^2 + 53xy + 63y^2$

6) $8x^2 - 2xy - 15y^2$

7) $4x^2 - 13xy + 9y^2$

8) $6x^2 + 23xy + 10y^2$

9) $5x^2 + 37xy - 24y^2$

10) $7a^2 - 22ab - 24b^2$

11) $2x^2 - 21xy + 49y^2$

12) $3x^2 + 8xy + 5y^2$

13) $55a^4 - 400a^3 + 420a^2$

14) $66p^3 + 474p^2 + 84p$

15) $39x^4 - 330x^3 - 189x^2$

16) $5n^2 - 36n + 14$

17) $3yx^2 + 10y^2x + 70y^3$

18) $2x^2 + 5xy - 12y^2$

19) $5x^2 - 22xy - 15y^2$

20) $2u^2 - 29uv + 39v^2$

21) $14u^2 + 47uv - 30v^2$

22) $36a^2b^2 - 92ab^3 + 40b^4$

23) $14a^2 + 149ab + 90b^2$

24) $9m^2 - 98mn - 120n^2$

25) $-12n^4 + 20n^3 - 3n^2$

26) $24x^4 - 123x^3 + 108x^2$

27) $-72p^3 + 688p^2 - 896p$

28) $-14n^2 - 43n + 90$

29) $28a^2 + 218ab + 234b^2$

30) $80x^2k + 424xky - 624y^2k$

31) $18x^2 - 144xy + 160y^2$

32) $6u^2 - 67uv + 152v^2$

Factoring All Mixed-Up

Now that you've learned the main methods and patterns of factoring, we will mix them all up and practice all the methods. Here are the key steps to factoring:

Step 1: Look for a **GCF**

Step 2: If it has **two** terms, look for a **difference of squares** and factor:

$$a^2 - b^2 = (a + b)(a - b)$$

Step 3: If it has **two** terms, look for a **sum of squares** and it cannot factor anymore:

$$a^2 + b^2 \text{ is prime.}$$

Step 4: If it has **two** terms, look for a **sum/difference of cubes**, use SOAP and factor:

$$a^3 \pm b^3 = (a \pm b)(a^2 \mp ab + b^2)$$

Step 5: If it has **three** terms, look for a **perfect square trinomial** and factor:

$$a^2 \pm 2ab + b^2 = (a \pm b)^2$$

Step 6: If it has **three** terms in the form $x^2 + bx + c$ look for factors of c that add up to b, make a table, **unFOIL**: $x^2 + bx + c = (x + m)(x + n)$

Step 7: If it has **three** terms in the form $ax^2 + bx + c$ look for factors of ac that add up to b, **grouping**: $ax^2 + bx + c = (px + m)(qx + n)$

Step 8: If it has **four** terms in the form use factoring by **grouping.**

As we get to more complicated problems, you may have to do more than one or two steps and some steps may need to be repeated.

Example 1: Factor $125x^2 - 200x + 80$

Solution: First look for a GCF, notice that 5 is the GCF so factor it out.

$$125x^2 - 200x + 80 = 5(25x^2 - 40x + 16)$$

Look at polynomial in the parentheses, maybe it can be factored down more. It is a trinomial with 3 terms, so we go to step 5. Check if it is a perfect square trinomial, and it is:

$$125x^2 - 200x + 80 = 5(25x^2 - 40x + 16) = 5(5x - 4)(5x - 4) \text{ or } 5(5x - 4)^2$$

One you get it broken down to monomial factors, linear factors (factors of the form px+m), and other prime factors you know you're done.

Example 2: Factor $8x^3 + 125$

Solution: First look for a GCF, but there is none for 8 and 125 (besides 1 of course). Notice it is a binomial with 2 terms. The terms are not squares, but they are cubes, so we have a sum of cubes and can factor:

$$8x^3 + 125 = (2x + 5)(4x^2 - 10x + 25)$$

You may wonder if the $4x^2 - 10x + 25$ can be factored again but recall this is not the case (unless it is a higher degree/exponent than 2).

Example 3: Factor $6b^3 + 24b^2 + 10b + 40$

Solution: First step to factoring, GCF which is 2 here:

$$6b^3 + 24b^2 + 10b + 40 = 2(3b^3 + 12b^2 + 5b + 20)$$

Running through the steps, we get to 4 terms factor by grouping, try without rearranging first:

$6b^3 + 24b^2 + 10b + 40 = 2((3b^3 + 12b^2) + (5b + 20))$. Let's just remember the GCF 2 and focus on the grouping:

$$(3b^3 + 12b^2) + (5b + 20) = 3b^2(b + 4) + 5(b + 4) = (b + 4)(3b^2 + 5).$$

Don't forget the GCF:

$$6b^3 + 24b^2 + 10b + 40 = 2(b + 4)(3b^2 + 5)$$

Example 4: Factor $4k^2 - 3k - 9$

Solution: There's no GCF and this has 3 terms. It cannot be a perfect square trinomial because the last sign is negative and if we take the product of the square roots of the first and last term then double it, $2k \cdot 3 \cdot 2 = 12$, it would not match the middle coefficient of -3. It seems to be a

general trinomial of the form $ax^2 + bx + c$. Search for factors of ac that add up to b.

Multiply to 4 x −9 = −36	Sum to −3
1 x −36	−35
2 x −18	−16
3 x −12	−9
4 x −9	−5
6 x −6	0
9 x −4	5

If the factors of -36 could add up to -3 then it would be between these two entries. Therefore, the polynomial is not factorable!

At this point, we get repetition, just the opposite of the numbers above

$4k^2 - 3k - 9$ is not factorable!

Practice Factoring All Mixed-Up

Factor each completely.

1) $294x^2 - 600$ $GCF(294, 600) = 6$

$6(49x^2 - 100)$

$6(7x + 10)(7x - 10)$

2) $224b^3 - 280b^2 + 192b - 240$

3) $9m^2 + 42m + 49$

4) $x^2 - 13xy + 36y^2$

5) $28pu + 18qv + 24pv + 21qu$

6) $49x^2 + 121$

7) $11x^2 - 37xy - 28y^2$

8) $12x^2 - 40xy - 63y^2$

9) $64u^3 + 27$

10) $64yx^3 - 343y^4$

11) $336ab - 240am - 294mb + 210m^2$

12) $2x^2 - 72y^2$

13) $6x^2 - 13x + 90$

14) $4vu^3 + 108v^4$

15) $324yx^2 + 576y^2x + 256y^3$

16) $9b^2 - 34b - 8$

17) $8x^3 - 27y^3$

18) $45u^3 - 430u^2v + 225v^2u$

19) $1200x^3 - 2028y^2x$

20) $64x^3 + 27y^3$

21) $36a^2b^5z + 84a^2b^5h + 42b^6az + 98b^6ah$

22) $100x^2 + 60xy + 9y^2$

23) $196u^2 + 400v^2$

24) $256yx^3 - 108y^4$

25) $8x^2 - 41xy + 5y^2$

26) $27u^3 - 8$

27) $72hm^2z - 16h^2n + 96h^2m^2 - 12hnz$

28) $x^3 + 8x^2y + 15y^2x$

29) $64m^3 + 27$

30) $-112a^2n + 328an + 224n$

31) $24uv + 12u - 9u^2 - 32v$

32) $9r^2 + 4r + 45$

33) $4x^2y + 80xy^2 + 384y^3$

34) $25m^2 + 70mn + 49n^2$

35) $3ba^3 - 192b^4$

36) $11n^3 - 6n^2 - 42n$

37) $14xy + 7x - 6vy - 3v$

38) $27x^3 + 343y^3$

39) $14a^2 + 33ab + 18b^2$

40) $20x^2 - 1805y^2$

Factoring Binomials Higher Degree

Factoring difference of squares and sum/difference of cubes also works with higher degree binomials. Factoring difference of squares pattern can work anytime the exponents are even:

$$a^{2n} - b^{2n} = (a^n + b^n)(a^n - b^n)$$

Example 1: Factor $(x^4 - y^4)$

Solution: Here the exponents are the same even degree with a minus sign between the terms.

Apply the difference of squares:

$(x^4 - y^4) = (x^2 + y^2)(x^2 - y^2)$ difference of squares can be used again:

$(x^4 - y^4) = (x^2 + y^2)(x^2 - y^2) = (x^2 + y^2)(x + y)(x - y)$

Factoring Binomials Higher Degree

Factoring sum/difference of cubes pattern can work anytime the exponents are multiples of three:

$$a^{3n} \pm b^{3n} = (a^n \pm b^n)(a^{2n} \mp a^n b^n + b^{2n})$$

Example 2: Factor $(x^9 + y^9)$

Solution: Here the exponents are the same multiples of three with a plus sign between the terms. Apply the sum of cubes:

$(x^9 + y^9) = (x^3 + y^3)(x^6 - x^3 y^3 + y^6)$ where the first factor can again be factored as a sum of cubes:

$$(x^9 + y^9) = (x^3 + y^3)(x^6 - x^3 y^3 + y^6) = (x + y)(x^2 - xy + y^2)(x^6 - x^3 y^3 + y^6)$$

Exponents that are multiples of 6 are both squares and cubes. If there is a minus sign you could do difference of squares or difference of cubes. So which method should you start with? Look at the next example.

Example 3: Factor $x^6 - y^6$

Solution: First try factoring as difference of cubes:

$$x^6 - y^6 = (x^2 - y^2)(x^4 + x^2 y^2 + y^4) = (x + y)(x - y)(x^4 + x^2 y^2 + y^2)$$

We have three factors, and it appears that $(x^4 + x^2 y^2 + y^2)$ cannot be factored anymore by any method discussed.

But look what happens if we try factoring as difference of squares:

$$x^6 - y^6 = (x^3 - y^3)(x^3 + y^3) = (x - y)(x^2 + xy + y^2)(x + y)(x^2 - xy + y^2)$$

Now we have four factors and matching up the factors, we have that:

$(x^4 + x^2y^2 + y^2)$ actually factors as:

$(x^4 + x^2y^2 + y^2) = (x^2 + xy + y^2)(x^2 - xy + y^2)$

From the previous example, for difference of exponents that are multiples of 6, you should start as a difference of squares. For sums, you will have to factor as a sum of cubes, since sum of squares generally don't factor. One exception of sum of squares that factors is the following:

$x^{10} + y^{10} = (x^2 + y^2)(x^8 - x^6y^2 + x^4y^4 - x^2y^6 + y^8)$

Practice Factoring Binomials Higher Degree

Factor each completely.

1) $x^4 - 1$

$(x^2 + 1)(x^2 - 1)$
$\qquad\qquad /\backslash$
$(x^2 + 1)(x + 1)(x - 1)$

2) $m^4 - 81$

3) $b^4 - 16$

4) $x^4 + 1$

5) $x^4 - 256$

6) $9k^4 - 25$

7) $x^6 - 16$

8) $r^6 - 4$

9) $4x^6 - 25$

10) $9x^6 - 25$

11) $125u^6 + 8$

12) $27u^6 + 8$

13) $64x^6 - 125$

14) $8x^6 - 125$

15) $u^6 - 1$

16) $x^6 - 64$

17) $x^6 - 125$

18) $u^6 + 27$

19) $x^8 - 1$

20) $x^8 + 1$

21) $x^9 - 1$

22) $x^9 + 1$

23) $x^{10} - 1$

24) $x^{12} - 1$

Factoring Quadratic Form

How can you factor $m^4 + 17m^2 + 72$? Upon inspection it looks like a trinomial similar to the form we have factored before: $ax^2 + bx + c$. The exponents are different but follow a pattern:

$ax^{2k} + bx^k + c$ which can be factored like $(px^k + m)(qx^k + n)$

Factoring Quadratic Form

Polynomials of this form are in quadratic form. For $m^4 + 17m^2 + 72$, we can make a substitution

$x = m^2$, then we can rewrite as $x^2 + 17x + 72$. Factor as usual:

$x^2 + 17x + 72 = (x + 8)(x + 9)$

Rewritten back in terms of m, it looks like:

$m^4 + 17m^2 + 72 = (m^2 + 8)(m^2 + 9)$

Notice that the degree of the binomials is half that of the trinomial.

Example 1: Factor $x^4 - 7x^2 - 18$

Solution: This is in quadratic form because it is a trinomial where the variable in the first term is

raised to twice the power of what it is raised to in the middle term. The last term is constant.

$x^4 - 7x^2 - 18 = (x^2 - 9)(x^2 + 2)$

Check these factors to see if any can be broken down further. We have a difference of squares:

$x^4 - 7x^2 - 18 = (x^2 - 9)(x^2 + 2) = (x + 3)(x - 3)(x^2 + 2)$

Example 2: Factor $x^6 - 9x^3 - 8$

Solution: Notice the first exponent is twice the second exponent and the last term is a constant.

We have another quadratic form. $x^6 - 9x^3 - 8 = (x^3 - 1)(x^3 - 8)$

We can factor both again as a difference of cubes:

$x^6 - 9x^3 - 8 = (x^3 - 1)(x^3 - 8) = (x - 1)(x^2 + x + 1)(x - 2)(x^2 + 2x + 4)$

Example 3: Factor $x^8 - x^4 - 12$

Solution: $x^8 - x^4 - 12 = (x^4 - 4)(x^2 + 3) = (x^2 - 2)(x^2 + 2)(x^2 + 3)$

Practice Factoring Quadratic Form

Factor each completely.

1) $u^4 - 5u^2 + 4$

$(u^2 - 1)(u^2 - 4)$

$(u+1)(u-1)(u+2)(u-2)$

2) $m^4 - 10m^2 + 25$

3) $u^4 - 7u^2 + 12$

4) $x^4 + 4x^2 - 96$

5) $x^4 - 4x^2y^2 - 140y^4$

6) $x^4 + 5x^2y^2 - 36y^4$

7) $x^4 + 12x^2y^2 + 20y^4$

8) $x^4 - 24x^2y^2 + 143y^4$

9) $x^5 - 15x^3y^2 + 56xy^4$

10) $4yx^4 - 52y^3x^2 + 160y^5$

11) $x^4 + 2x^2y^2 + 10y^4$

12) $x^4 + 16x^2y^2 + 60y^4$

13) $3yx^4 + 26y^3x^2 - 169y^5$

14) $11x^4 - 30x^2y^2 - 56y^4$

15) $24x^5 + 200x^3y^2 - 144xy^4$

16) $7x^4 - 11x^2y^2 - 6y^4$

17) $4a^4 + 4a^2 + 5$

18) $16x^4 - 16x^2 - 45$

19) $90x^5 - 312x^3 + 168x$

20) $24x^5 - 62x^3 + 35x$

21) $x^6 + 22x^3 + 112$

22) $m^8 + 19m^4 + 70$

23) $x^8 - 10x^4 + 24$

24) $x^6 - 7x^3 - 8$

25) $3x^8 + 10x^4y^4 - 13y^8$

26) $13x^{10}y^2 + 16x^6y^6 + 3x^2y^{10}$

27) $7x^6 + 104x^3y^3 + 169y^6$

28) $13x^6 - 16x^3y^3 + 14y^6$

29) $9yx^6 + 9y^7$

30) $8x^8 - 30x^4y^4 + 7y^8$

31) $9x^8 + 10y^8$

32) $224nm^8 - 792n^5m^4 + 648n^9$

Solving Quadratic Equations by Factoring

One reason that factoring is so important is that it allows us to solve a variety of problems. One of the typical problems in algebra is solving quadratic equations. A quadratic equation is an equation of degree 2 (highest exponent is 2) that can be written in standard form as: $ax^2 + bx + c = 0$. Quadratic equations arise in many different fields including physics problems dealing with projectile motion. There are several ways to solve quadratic equations including: factoring, extracting square roots, completing the square, quadratic formula, and graphing.

Solving Quadratic Equations by Factoring

If the product of two numbers is equal to zero, then what can we say about the two numbers? One or both numbers must equal zero. This is known as the **zero product property**:

$$If\ pq = 0, then\ p = 0\ or\ q = 0$$

We can solve a quadratic by setting it to zero, try to factor the quadratic and setting each factor to zero. **A quadratic can be factored if and only if it has "nice" rational solutions.**

Example 1: Solve $(x - 5)(4x - 3) = 0$

Solution: Since this is already factored, set each factor to zero and solve for each:

$$x - 5 = 0\ or\ 4x - 3 = 0$$

$$x = 5\ or\ x = \frac{3}{4}$$

The zero product property only works if a product is set to zero:

Example 2: Solve $n^2 - 3n = 28$

Solution: First set to zero:

$$n^2 - 3n - 28 = 0$$

Factor and set to zero:

$$(n - 7)(n + 4) = 0$$

$$n - 7 = 0\ or\ n + 4 = 0$$

$$n = 7\ or\ n = -4$$

Example 3: Solve $14x^2 - 34x + 16 = -6x^2 + 4$

Solution: Set to zero and combine like terms to get the equation into standard form ax² + bx + c = 0:

$20x^2 - 34x + 12 = 0$

Factor by taking out a GCF first:

$2(10x^2 - 17x + 6) = 0$

$2(2x - 1)(5x - 6) = 0$

The GCF of 2 does not give any solutions. Set the other factors to zero:

$2x - 1 = 0 \ or \ 5x - 6 = 0$

$x = \dfrac{1}{2} \ or \ x = \dfrac{6}{5}$

Practice Solving Quadratic Equations by Factoring

Solve each equation by factoring.

1) $(12a - 5)(a - 12) = 0$

$12a - 5 = 0 \quad or \quad a - 12 = 0$
$12a = 5$
$a = \frac{5}{12} \quad or \quad a = 12$

2) $p(p - 2) = 0$

3) $(x - 5)(x + 10) = 0$

4) $(n + 12)(n + 11) = 0$

5) $r^2 + 6r + 8 = 0$

6) $n^2 - 21n + 108 = 0$

7) $v^2 + 6v = 0$

8) $a^2 - 6a + 5 = 0$

9) $a^2 - 2a = 48$

10) $r^2 + 20 = -9r$

11) $m^2 + 84 = -19m$

12) $b^2 = -54 - 15b$

13) $n^2 + 9n - 39 = -3$

14) $n^2 + 48 = 14n$

15) $m^2 + 8m + 24 = -6 - 5m$

16) $p^2 - 4p - 69 = -11p - 9$

17) $8n^2 - 4n - 35 = 6n^2 - 5$

18) $7x^2 - 31x - 45 = 4x - 3$

19) $b^2 - 8b + 9 = -6$

20) $x^2 - 10x + 7 = 7 - 6x$

21) $5m^2 + 13m - 28 = 0$

22) $12v^2 - 7v = 0$

23) $77v^2 - 97v + 30 = 0$

24) $7x^2 - 3x - 4 = 0$

25) $7k^2 = 12 + 5k$

26) $11b^2 = 12 - b$

27) $5a^2 = 49 - 28a$

28) $9n^2 - 49 = 56n$

29) $70n^2 + 330n + 210 = 10$

30) $21k^2 + 108k - 661 = 11 - 3k^2$

31) $7n^2 + 47n + 61 = 1$

32) $100x^2 - 7x - 590 = 3x + 10$

Simplifying Radicals

A radical is a root, such as a square root $\sqrt{a}$, cube root $\sqrt[3]{a}$ or higher. As we solve quadratic equations and higher degree polynomial equations, we will run into solutions that are not rational (fractions) but may be irrational and involve radicals to solve.

What is a radical anyways? A radical is the inverse operation or reverse operation of raising a number to a power. Just like the inverse of addition is subtraction, and the inverse of multiplication is division, well the inverse of squaring a number is taking the square root and the inverse of cubing a number is, you guessed it, the cube root. If $a^n = b$ then $\sqrt[n]{b} = a$. These roots are radical!

INDEX

$$\sqrt[n]{a} \longleftarrow \textbf{RADICAND}$$

Because squaring and square rooting are inverses, we can easily solve problems like

$$\sqrt{1379}\sqrt{1379} = \left(\sqrt{1379}\right)^2 = 1379$$

The properties of exponents can be rewritten to apply for radicals:

Properties of Radicals

Property Name	Notation	Explanation	Numeric Example	Algebraic Example
Nth Root Exponent	$\sqrt[n]{a} = a^{\frac{1}{n}}$	nth root is the same a raising to the 1/n power	$9^{\frac{1}{2}} = \sqrt{9} = 3$	$\sqrt[3]{8x} = 2x^{\frac{1}{3}}$
Root of a product	$\sqrt[n]{ab} = \sqrt[n]{a}\sqrt[n]{b}$	Radical of a product is the product of the radicals	$\sqrt{20} = \sqrt{4}\sqrt{5} = 2\sqrt{5}$	$\sqrt{18x^3} = \sqrt{9x^2}\sqrt{2x}$ $= 3x\sqrt{2x}$
Root of quotient	$\sqrt[n]{\frac{a}{b}} = \frac{\sqrt[n]{a}}{\sqrt[n]{b}}$	Radical of a quotient is the quotient of the radicals	$\sqrt{\frac{4}{9}} = \frac{\sqrt{4}}{\sqrt{9}} = \frac{2}{3}$	$\sqrt[3]{\frac{x^6}{8}} = \frac{\sqrt[3]{x^6}}{\sqrt[3]{8}} = \frac{x^2}{2}$
Fractional Exponents	$a^{\frac{m}{n}} = \sqrt[n]{a^m}$ $= (\sqrt[n]{a})^m$	Rational exponents can be written in radical form	$8^{\frac{2}{3}} = (\sqrt[3]{8})^2 = 2^2 = 4$	$x^{\frac{5}{2}} = (\sqrt{x})^5$
Odd Root to Power	$(\sqrt[n]{a})^n = \sqrt[n]{a^n}$ $= a,$ $n\ is\ odd$	A nth root raised to the nth power is equal to the number	$(\sqrt[3]{127})^3 = 127$	$(\sqrt[7]{x})^7 = x$
Even Root to Power	$\sqrt[n]{a^n} = |a|,$ $n\ is\ even$	A nth root raised to the nth power is equal to the absolute value of the number	$\sqrt{(-5)^2} = |-5| = 5,$	$\sqrt[6]{x^6} = |x|,$
Rationalizing a Root	$\frac{a}{\sqrt{b}} = \frac{a}{\sqrt{b}} \cdot \frac{\sqrt{b}}{\sqrt{b}}$ $= \frac{a\sqrt{b}}{b}$	Use this to simplify expressions with a radical in the denominator	$\frac{2}{\sqrt{3}} = \frac{2}{\sqrt{3}} \cdot \frac{\sqrt{3}}{\sqrt{3}} = \frac{2\sqrt{3}}{3}$	$\frac{xy}{\sqrt{x}} = \frac{xy}{\sqrt{x}} \cdot \frac{\sqrt{x}}{\sqrt{x}} = y\sqrt{x}$

We are going to focus on simplifying radicals, especially square roots since we will need that to solve quadratic equations and other problems later. We will break up or factor a radical, using this property to simplify radicals:

$$\sqrt[n]{ab} = \sqrt[n]{a}\sqrt[n]{b}$$

For square roots we want the biggest square root that divides into the radicand, and we will pull that out to the front so we can take the square root of it. We want to make the number that is under the radical to be as small as possible.

The following radicals are all equivalent $\sqrt{80} = 2\sqrt{20} = 4\sqrt{5}$, but the simplified radical is $4\sqrt{5}$ since the number under the square root is smallest.

Simplifying Radicals

Example 1: Simplify $\sqrt{48}$

Solution: What is the biggest square number that divides evenly into 48? Well 49 is too big, 36 doesn't work, 25 doesn't work either... 16 works! Factor 48 into 16 and 3, putting the square in front:

$$\sqrt{48} = \sqrt{16}\sqrt{3} = 4\sqrt{3}$$

Notice that 4 also goes into 48, but you should find the biggest square number that works.

Example 2: Simplify $\sqrt{28}$

Solution: $\sqrt{28} = \sqrt{4}\sqrt{7} = 2\sqrt{7}$

Example 3: Simplify $4\sqrt{18}$

Solution: $4\sqrt{18} = 4\sqrt{9}\sqrt{2} = 4 \cdot 3\sqrt{2} = 12\sqrt{2}$

For cube roots, factor the radicand into the biggest cube and the remaining factor:

Example 4: Simplify $\sqrt[3]{375}$

Solution: $\sqrt[3]{375} = \sqrt[3]{125}\sqrt[3]{3} = 5\sqrt[3]{3}$

You can also simplify cube roots of negative numbers. Always take the negative outside to the front of the cube root. Never leave the negative in the cube root:

Example 5: Simplify $\sqrt[3]{-320}$

Solution: $\sqrt[3]{-320} = \sqrt[3]{-64}\sqrt[3]{5} = -4\sqrt[3]{5}$

Practice Simplifying Radicals

Simplify.

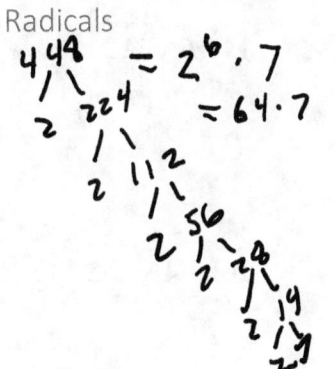

1) $\sqrt{448}$

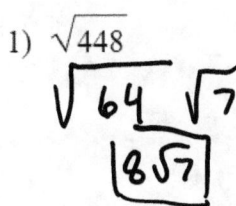

2) $\sqrt{32}$

3) $\sqrt{75}$

4) $\sqrt{128}$

5) $\sqrt{196}$

6) $\sqrt{108}$

7) $\sqrt{175}$

8) $\sqrt{45}$

9) $\sqrt{294}$

10) $\sqrt{343}$

11) $3\sqrt{72}$

12) $3\sqrt{128}$

13) $-4\sqrt{24}$

14) $6\sqrt{112}$

15) $7\sqrt{96}$

16) $-4\sqrt{252}$

17) $-4\sqrt{112}$

18) $4\sqrt{128}$

19) $\sqrt[6]{192}$

20) $\sqrt[3]{448}$

21) $\sqrt[3]{32}$

22) $\sqrt[3]{750}$

23) $\sqrt[3]{16}$

24) $\sqrt[3]{48}$

25) $\sqrt[3]{-135}$

26) $\sqrt[3]{-512}$

27) $\sqrt[3]{81}$

28) $\sqrt[4]{96}$

29) $\sqrt{2535}$

30) $\sqrt{675}$

31) $\sqrt{637}$

32) $\sqrt{2025}$

Complex Numbers

There are some equations that cannot be solved with real solutions. One such example is the equation $x^2 + 1 = 0$ which then becomes $x^2 = -1$. But what number squared can ever be a negative? There is no real number that squared becomes negative. In order to solve problems like this, mathematicians developed a number defined as:

$i = \sqrt{-1}$

Complex Numbers

This allows the simplification of all negative roots, breaking up the square root as previously done.

Example 1: Simplify $\sqrt{-16}$

Solution: $\sqrt{-16} = \sqrt{16}\sqrt{-1} = 4i$

Example 2: Simplify $\sqrt{-50}$

Solution: $\sqrt{-50} = \sqrt{25}\sqrt{2}\sqrt{-1} = 5\sqrt{2}i$ or sometimes written as $5i\sqrt{2}$

Numbers with an *i* like this are called **imaginary numbers.** Sometimes when students hear that these are imaginary numbers, they think this is not "real" math and just made up on the spot. But all math is an abstraction, all numbers are conceptual. Mathematics works because it is logically consistent and built up from previous fundamental ideas. Imaginary numbers are used in "real world" applications especially in electrical engineering and quantum mechanics.

A real number added or subtracted to an imaginary number gives a **complex number** in the form *a+bi*. Operations can be performed with imaginary and complex numbers in the same way as with **real numbers**. With these new numbers, we will be able to find solutions to any quadratic equation!

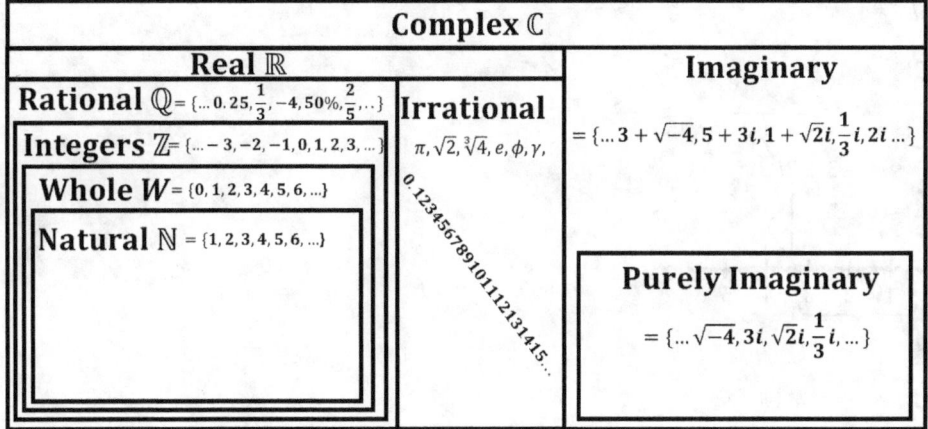

Just as the real numbers can be plotted on the real number line, the imaginary numbers can be plotted on the imaginary number line. These lines only intersect at zero, and we consider them to be perpendicular forming the two axes of the complex plane.

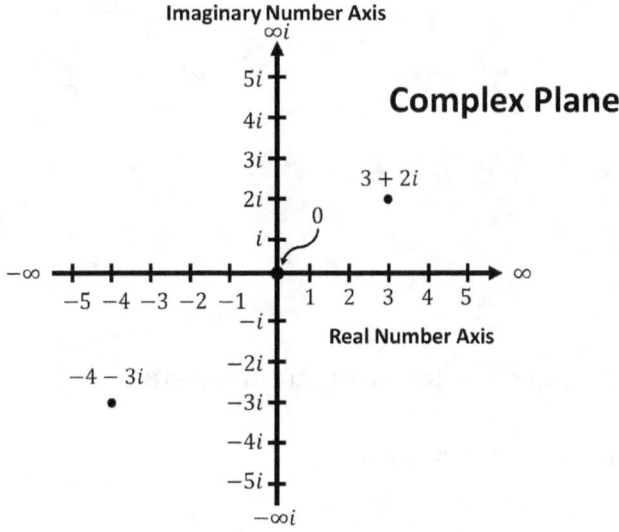

Absolute Value is distance from zero. To find the absolute value of a complex number we need to find how far that number is from the origin on the graph. If the complex number is purely imaginary or real, then you can count how many spaces it is from zero. But if the complex number has both a real and imaginary part, then use the Pythagorean Theorem to find the distance. Only use the positive horizontal and vertical lengths in the Pythagorean Theorem.

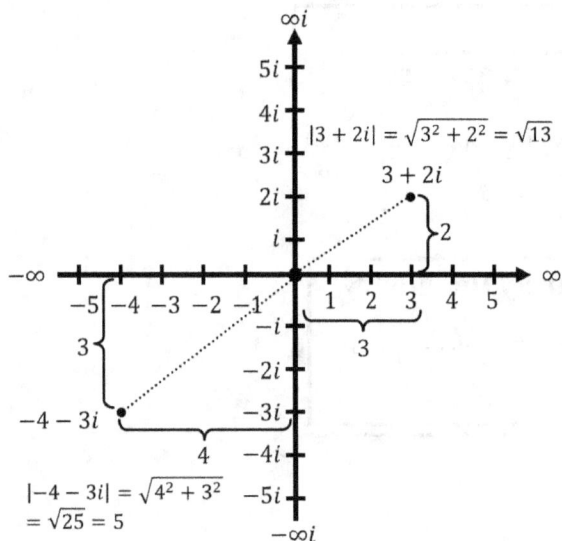

We can find powers of i by using the definition of $i = \sqrt{-1}$ and properties of exponents.

$$i = \sqrt{-1}$$

$$i^2 = \left(\sqrt{-1}\right)^2 = -1$$

$$i^3 = i \cdot i^2 = i \cdot -1 = -i$$

$$i^4 = i^2 \cdot i^2 = -1 \cdot -1 = 1$$

$i^5 = i \cdot i^4 = i \cdot 1 = i$ And now we are back at the beginning and the pattern repeats $i^5 = i$

$i^6 = i^2 \cdot i^2 \cdot i^2 = -1 \cdot -1 \cdot -1 = -1$ This is the same as previously $i^6 = i^2$

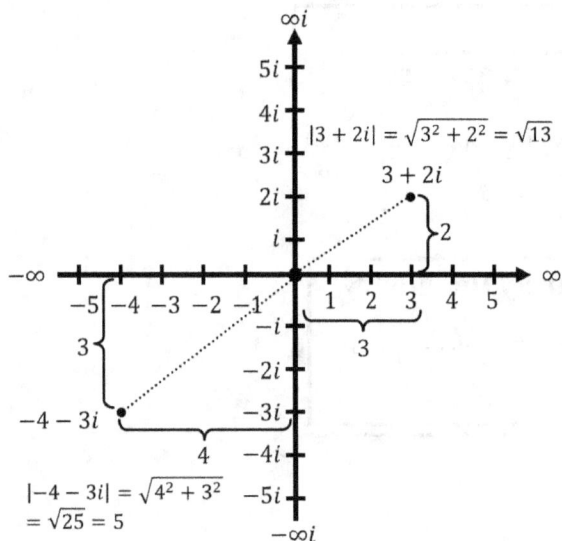

We can use this pattern to find higher powers of i, by taking the exponent, dividing it by 4 and finding the remainder. The remainder will tell us where we are in the cycle.

Example 3: Simplify i^{500}

Solution:

$$\begin{array}{r} 125 \; R0 \\ 4\overline{)500} \\ \underline{4} \\ 10 \\ \underline{8} \\ 20 \end{array}$$

A remainder of zero is the same as $i^4 = 1$

$i^{500} = 1$

Example 4: Simplify i^{270}

Solution:

$$\begin{array}{r} 67 \; R2 \\ 4\overline{)270} \\ \underline{24} \\ 30 \\ \underline{28} \\ 2 \end{array}$$

A remainder of two is the same as $i^2 = -1$

$i^{270} = -1$

Add or subtract complex numbers by combining like terms, adding the real terms to the real terms and the imaginary terms to the imaginary terms. For subtraction, distribute the negative inside the parentheses.

$$(a + bi) + (c + di) = (a + c) + (b + d)i \qquad (a + bi) - (c + di) = (a - c) + (b - d)i$$

Complex Numbers

Example 5: Simplify $(2 - 6i) + (4 + 3i)$

Solution: $(2 - 6i) + (4 + 3i) = 2 + 4 - 6i + 3i = 6 - 3i$

Example 6: Simplify $(2 - 6i) - (4 + 3i)$

Solution: $(2 - 6i) - (4 + 3i) = 2 - 4 - 6i - 3i = -2 - 9i$

To multiply complex numbers simply distribute or use the FOIL method as used for multiplying binomials. Remember that for the terms that contain i^2 that part will become a real opposite term $i^2 = -1$

Example 7: Simplify $(2 - 6i)(4 + 3i)$

Solution: $(2 - 6i)(4 + 3i) = 8 + 6i - 24i - 18i^2 = 8 + 18 + 6i - 24i = 26 - 18i$

Example 8: Simplify $(2 - 3i)^2$

Solution: $(2 - 3i)^2 = (2 - 3i)(2 - 3i) = 4 - 6i - 6i + 9i^2 = 4 - 9 - 6i - 6i = -5 - 12i$

When dividing complex and imaginary numbers, we cannot leave i in the denominator. When dividing by a purely imaginary number, multiply the top and bottom of the fraction by i.

Example 9: Simplify $\frac{4 - 6i}{2i}$

Solution: $\frac{4 - 6i}{2i} = \frac{4 - 6i}{2i} \cdot \frac{i}{i} = \frac{4i - 6i^2}{2i^2} = \frac{4i + 6}{-2} = -3 - 2i$

Dividing by a complex binomial requires multiplying by the conjugate of the denominator.

The conjugate changes the sign of the last term.

Conjugate of $a + bi = a - bi$

Multiplying by the conjugate works because we end up with a difference of squares which removes the imaginary terms.

$$(a + bi)(a - bi) = a^2 - abi + abi - bi^2 = a^2 + b^2$$

This also allows one to "factor" a sum of squares using complex numbers.

Example 9: Simplify $\frac{4-6i}{1+2i}$

Solution: $\frac{4-6i}{1+2i} = \frac{4-6i}{1+2i} \cdot \frac{1-2i}{1-2i} = \frac{4-8i-6i+12i^2}{1-4i^2} = \frac{4-12-8i-6i}{1+4} = \frac{-8-14i}{5}$

$$\frac{4-6i}{1+2i} = \frac{-8-14i}{5} \ or \ -\frac{8}{5} - \frac{14}{5}i$$

Practice with Complex Numbers

Graph each number in the complex plane.

1) $-2i$

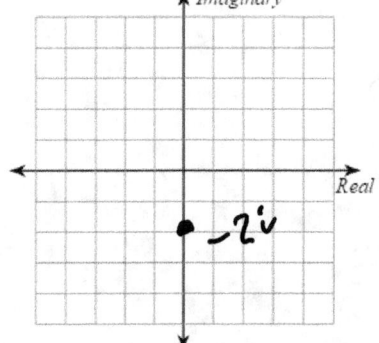

2) $3 - 4i$

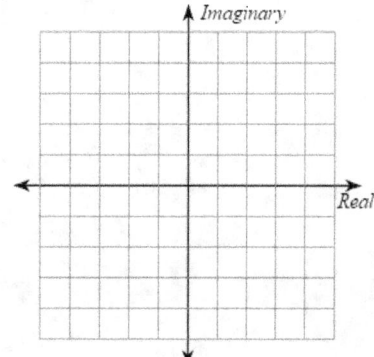

3) $-1 + 3i$

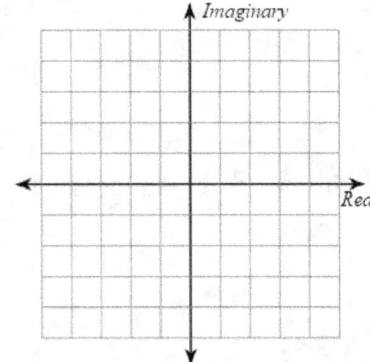

4) $-4 - 3i$

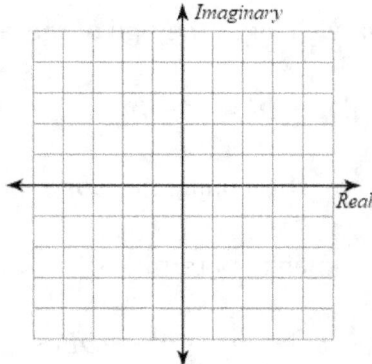

Find the absolute value of each complex number.

5) $\left| i \right|$

6) $\left| -5i \right|$

7) $\left| -2 + i \right|$

8) $\left| 2 + 2i \right|$

Simplify

9) $\sqrt{-81}$

10) $\sqrt{-36}$

11) $\sqrt{-96}$

12) $\sqrt{-20}$

13) $\sqrt{-720}$

14) $\sqrt{-18}$

15) $\sqrt{-75}$

16) $\sqrt{-192}$

13) $\sqrt{-720}$

14) $\sqrt{-18}$

15) $\sqrt{-75}$

16) $\sqrt{-192}$

17) $6\sqrt{-12}$

18) $-7\sqrt{-48}$

19) $-5i\sqrt{-98}$

20) $5\sqrt{-100}$

Simplify the following:

21) i^{25}

22) i^{90}

23) i^{35}

24) i^{80}

25) i^{120}

26) i^{190}

27) i^{235}

28) i^{125}

29) $(8 + 6i) + (-6 + 4i)$

30) $(3 + 8i) - (-8 + i)$

31) $(3 - 4i) - (6 + 5i)$

32) $(-1 + 4i) + (6 + i)$

33) $(-6 - i)(5 - 8i)$

34) $(7 - 5i)(-8 - 3i)$

35) $(3 + 5i)(3 + 2i)$

36) $(1 + 2i)(4 - 3i)$

37) $(5 + 3i)^2$

38) $(6 + i)^2$

39) $(5 - 5i)^2$

40) $(-4 - 4i)^2$

Divide.

41) $\dfrac{-9 + 6i}{2i}$

42) $\dfrac{10 + 10i}{2i}$

43) $\dfrac{2 - 7i}{4 - 8i}$

44) $\dfrac{6i}{10 + 5i}$

45) $\dfrac{6i}{-6+3i}$

46) $\dfrac{-2-7i}{2+3i}$

47) $\dfrac{-3+7i}{9+8i}$

48) $\dfrac{8+4i}{2+4i}$

Solving Quadratics by Extracting Square Roots

To solve the equation $x^2 = 16$, you can set it to zero, factor and set each factor to zero.

$x^2 = 16 \Longrightarrow x^2 - 16 = 0 \Longrightarrow (x+4)(x-4) = 0 \Longrightarrow x = -4 \: or \: x = 4$

You can also simply take the square roots of both sides, but you need to take the positive and

negative roots of the number.

$x^2 = 16 \Longrightarrow x^2 = \pm\sqrt{16} \Longrightarrow x^2 = \pm 4 \Longrightarrow x = -4 \: or \: x = 4$

Some equations cannot be factored over the integers, but we can still use the method of taking

roots on both sides. Extracting square roots method works for quadratic equations that do not

have a linear term and can be written in the form: $ax^2 + c = 0$

Example 1: Solve $x^2 = 48$

Solution: Taking the positive and negative square root of both sides gives:

$x = \pm\sqrt{48} = \pm\sqrt{16}\sqrt{3} = \pm 4\sqrt{3}$

Solving Quadratics by Extracting Square Roots

Sometimes the square term is not by itself, so first isolate the square then take square roots:

Example 2: Solve $4x^2 + 10 = 190$

Solution: Solve by x^2 by subtracting 10 and dividing by 4:

$$4x^2 + 10 = 190 \implies 4x^2 = 180 \implies x^2 = 45$$

Now extract square roots and simplify the radical:

$$x^2 = 45 \implies x = \pm\sqrt{45} \implies x = \pm\sqrt{9}\sqrt{5} \implies x = \pm 3\sqrt{5}$$

We have two answers: $x = 3\sqrt{5} \ or \ x = -3\sqrt{5}$ but we can just write both at the same time $x = \pm 3\sqrt{5}$

Example 3: Solve $(x + 3)^2 = 36$

Solution: It may be tempting to square out the binomial and solve by factoring, and while that works, it makes the problem more difficult. This is already written as a perfect square equal to a number, so extracting square roots is the easiest way to proceed.

$$(x + 3)^2 = 36 \implies x + 3 = \pm\sqrt{36} \implies x + 3 = \pm 6 \implies x = -3 \pm 6$$

Since we can add and subtract the terms, this should be simplified to $x = -9 \ or \ x = 3$

Example 4: Solve $(x + 4)^2 = -8$

Solution: $(x + 4)^2 = -8 \implies x + 4 = \pm\sqrt{-8} \implies x + 4 = \pm\sqrt{4}\sqrt{2}\sqrt{-1} \implies x = -4 \pm 2\sqrt{2}i$

On this one we cannot add and subtract the terms, this cannot be simplified and is best to just write the answer as: $x = -4 \pm 2\sqrt{2}i$

Practice Solving Quadratics by Extracting Square Roots

Solve each equation by taking square roots.

1) $\sqrt{v^2} = \sqrt{49}$

$v = \pm 7$

2) $x^2 = 64$

3) $m^2 = 147$

4) $r^2 = 135$

5) $36x^2 + 5 = 69$

6) $49r^2 + 13 = 22$

7) $4a^2 + 11 = 731$

8) $14b^2 + 11 = 893$

9) $-4 - 5n^2 = -9$

10) $49m^2 + 4 = 29$

11) $7p^2 + 9 = 373$

12) $3v^2 - 5 = 97$

13) $(x + 4)^2 = 81$

14) $(x - 4)^2 = 36$

15) $(x + 6)^2 = 25$

16) $(x - 6)^2 = 49$

17) $(x - 3)^2 = 40$

18) $(x + 5)^2 = 72$

19) $(x - 4)^2 = 24$

20) $(x + 6)^2 = 75$

21) $(x + 4)^2 = -81$

22) $(x - 5)^2 = -64$

23) $(x - 6)^2 = -32$

24) $(x + 5)^2 = -72$

Solving Quadratics by Completing the Square

Quadratics can be solved by completing the square. When completing the square, we try to make the trinomial into a perfect square and then extract square roots to solve. We can represent polynomials visually with algebra tiles.

Look at the following pattern, how many small squares must be added to complete the square to make a large square:

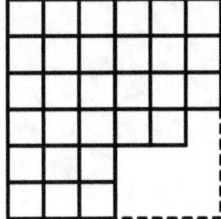

The figure above has 29 little squares, which is not a perfect square but if we add seven squares, we end up with a perfect square of 36.

Now imagine you're given an algebraic expression $x^2 + 4x + 1$ which could be represented visually as:

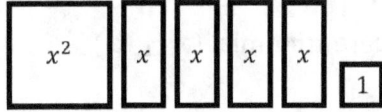

How could you arrange these tiles to form one large square?

Start with the x^2 tile in the top left, then use the x rectangles to the right and below. You need to split the x rectangles into two groups. Then use the unit 1 squares to fill in as much space as possible. The goal is to make a perfect square with the fewest unit tiles.

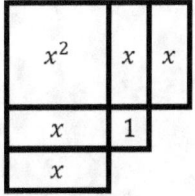

Solving Quadratics by Completing the Square

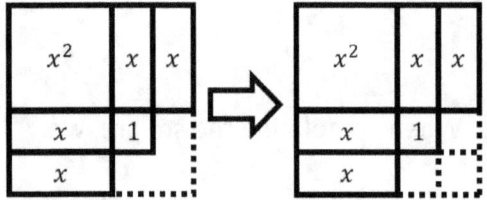

Visually adding 3 will make a perfect square.

Algebraically, if we add 3 to the original expression, we will be able to factor the quadratic as a perfect square trinomial:

$$x^2 + 4x + 1 + 3 = x^2 + 4x + 4 = (x + 2)^2$$

Steps to Completing the Square:

1) Put the quadratic into standard form $ax^2 + bx + c = 0$

2) If $a \neq 1$, then divide the entire equation by a

3) Move the constant to the other side

4) Now to complete the square, take half the coefficient of the x term and square it. Add this to both sides.

5) You should be able to factor the left side as a perfect square trinomial $(x \pm d)^2$

6) Solve for x by taking square roots

Example 1: Find the value that completes the square, then rewrite as a perfect square:

$$x^2 - 12x + \underline{\quad}$$

Solution: Since the leading coefficient is 1, take half of −12 and square it which is 36

$$x^2 - 12x + 36 = (x - 6)^2$$

Example 2: Solve by completing the square $x^2 + 6x - 59 = 0$

Solution: The quadratic is already in standard form with a leading coefficient of 1. Move the constant to the other side:

$x^2 + 6x = 59$

Focus on the left side to complete the square. Take half the coefficient of x and square it.

$\left(\frac{6}{2}\right)^2 = 9$

Add this number to both sides:

$x^2 + 6x + 9 = 59 + 9$

$x^2 + 6x + 9 = 68$

$(x + 3)^2 = 68$

$x + 3 = \pm\sqrt{68}$

$x = -3 \pm \sqrt{4}\sqrt{17}$

$x = -3 \pm 2\sqrt{17}$

Example 3: Solve by completing the square $9x^2 - 18x - 18 = 3$

Solution: Move the constant to the other side:

$9x^2 - 18x = 21$

The leading coefficient is not 1, so divide all terms by 9.

$x^2 - 2x = \frac{7}{3}$

Take half the coefficient of -2 and square it, adding to both sides:

$x^2 - 2x + 1 = \frac{7}{3} + 1$

$$(x + 1)^2 = \frac{10}{3} \Rightarrow x + 1 = \pm\sqrt{\frac{10}{3}} \Rightarrow x = -1 \pm \frac{\sqrt{10}}{\sqrt{3}} \Rightarrow -\frac{3}{3} \pm \frac{\sqrt{30}}{3} = \frac{-3 \pm \sqrt{30}}{3}$$

Practice Solving Quadratics by Completing the Square

Find the value that completes the square and then rewrite as a perfect square.

1) $x^2 + 10x +$ __25__ $= (x+5)^2$

$\left(\frac{10}{2}\right)^2 = 25$

2) $x^2 + 6x +$ ___

3) $x^2 + 38x +$ ___

4) $x^2 - 36x +$ ___

5) $x^2 - 26x +$ ___

6) $y^2 - 24y +$ ___

7) $n^2 - 28n +$ ___

8) $n^2 - 10n +$ ___

9) $n^2 - 11n + \underline{\quad}$

10) $x^2 - 9x + \underline{\quad}$

11) $x^2 - x + \underline{\quad}$

12) $n^2 + 5n + \underline{\quad}$

Solve each equation by completing the square.

13) $x^2 - 16x + 8 = 0$

14) $m^2 + 18m + 67 = 0$

15) $x^2 + 2x + 55 = 0$

16) $x^2 - 18x + 99 = 0$

17) $k^2 - 11k + 25 = 0$

18) $n^2 + 7n - 26 = 0$

19) $n^2 + 16n - 32 = 0$

20) $x^2 - 10x + 48 = 0$

21) $b^2 - 16b + 28 = 0$

22) $k^2 + 6k + 90 = 0$

23) $6x^2 = 5x^2 - 7 - 12x$

24) $-4x^2 - 23x = -39 - 5x^2 - 9x$

25) $2x^2 - 20x + 32 = 0$

26) $7a^2 + 14a - 56 = 0$

27) $4p^2 + 8p - 52 = 0$

28) $5a^2 - 20a - 10 = 0$

29) $x^2 + 3x - 62 = -3$

30) $a^2 - 9a + 22 = 9$

31) $x^2 + 5x - 76 = -10$

32) $r^2 + r - 39 = 8$

33) $4r^2 + 11r = -5r - 68$

34) $9b^2 - 11b - 67 = 7b$

35) $4x^2 - 117 = 8x$

36) $10v^2 + 21v + 106 = v$

37) $25n^2 - 19n + 9 = 12n^2 - 10n$

38) $11x^2 - 95 = 13 + 4x^2 - x$

39) $x^2 + 9x + 12 = 5$

40) $x^2 - 6x = 169 - 13x$

Solving Quadratics by Quadratic Formula

The quadratic formula is one of the most famous formulas in mathematics. It is derived from completing the square from the standard form of a quadratic formula. Here's how the formula is derived:

1. $ax^2 + bx + c = 0$

2. $ax^2 + bx = -c$

3. $x^2 + \dfrac{b}{a}x = -\dfrac{c}{a}$

4. $x^2 + \dfrac{b}{a}x + \left(\dfrac{b}{2a}\right)^2 = \left(\dfrac{b}{2a}\right)^2 - \dfrac{c}{a}$

5. $\left(x + \dfrac{b}{2a}\right)^2 = \left(\dfrac{b}{2a}\right)^2 - \dfrac{c}{a}$

6. $\left(x + \dfrac{b}{2a}\right)^2 = \dfrac{b^2}{4a^2} - \dfrac{c}{a}$

7. $\left(x + \dfrac{b}{2a}\right)^2 = \dfrac{b^2}{4a} - \dfrac{4ac}{4a^2}$

8. $\left(x + \dfrac{b}{2a}\right)^2 = \dfrac{b^2 - 4ac}{4a^2}$

9. $x + \dfrac{b}{2a} = \dfrac{\sqrt{b^2 - 4ac}}{\sqrt{4a^2}}$

10. $x = -\dfrac{b}{2a} \pm \dfrac{\sqrt{b^2 - 4ac}}{2a}$

11. $x = \dfrac{-b \pm \sqrt{b^2 - 4ac}}{2a}$

Quadratic Equation Standard Form: $ax^2 + bx + c = 0$

Quadratic Formula: $x = \dfrac{-b \pm \sqrt{b^2 - 4ac}}{2a}$

Steps to Solving by Quadratic Formula

1) Set the equation to zero and put into standard form $ax^2 + bx + c = 0$

2) If the leading coefficient a is negative, multiply each term by -1

   Now you have standard form with positive a

3) Identify the coefficients $a, b, and\ c$

4) Plug into the quadratic formula $x = \dfrac{-b \pm \sqrt{b^2 - 4ac}}{2a}$ Don't forget the $\pm$

Solving Quadratics by Quadratic Formula

5) Simplify the **discriminant** part which is $b^2 - 4ac$

6) Simplify the square root of $b^2 - 4ac$

   Solutions may be "nice" rational, irrational, or imaginary!

7) Simplify fractions, dividing every term by the GCF

Example 1: Solve $12x^2 - 12x - 5 = 0$

Solution: Equation is in standard form, set to zero with a positive leading coefficient. Identify

the coefficients a, b, and c. a = 12, b = –12, c = –5   Plug these into the formula:

$$x = \frac{-(-12) \pm \sqrt{(-12)^2 - 4(12)(-5)}}{2(12)}$$

Simplify the discriminant part $(-12)^2 - 4(12)(-5) = 384$

$$x = \frac{12 \pm \sqrt{384}}{24}$$

Simplify the square root:

$$x = \frac{12 \pm 8\sqrt{6}}{24}$$

Simplify the fraction by dividing every term by GCF, i.e., dividing 12, 8, and 24 by 4:

$$x = \frac{3 \pm 2\sqrt{6}}{6}$$

Example 2: Solve $-10x^2 + 12x - 8 = -x^2 + 4x - 6$

Solution: Move all the terms to the left and set to zero

$$-9x^2 + 8x - 2 = 0$$

Notice that the leading coefficient is negative. The quadratic formula still works with a negative value for a, but it's a lot easier if it's positive:

$$9x^2 - 8x + 2 = 0$$

Identify the coefficients a, b, and c. a = 9, b = –8, c = 2   Plug these into the formula:

$$x = \frac{-(-8) \pm \sqrt{(-8)^2 - 4(9)(2)}}{2(9)}$$

Simplify the discriminant part $(-8)^2 - 4(9)(2) = -8$

$$x = \frac{8 \pm \sqrt{-8}}{18}$$

Simplify the square root:

$$x = \frac{8 \pm 2\sqrt{2}i}{18}$$

Simplify the fraction by dividing every term by GCF, i.e., dividing 8, 2, and 12 by 2:

$$x = \frac{4 \pm \sqrt{2}i}{9}$$

Example 3: Solve $2n^2 + 4n - 96 = 0$

Solution: Even though the variable is not x, just think of n as x and use the formula. The leading coefficient is positive. Find the coefficients: a = 2, b = 4, c = –96. Plug these into the formula:

$$x = \frac{-(4) \pm \sqrt{(4)^2 - 4(2)(-96)}}{2(2)}$$

Simplify the discriminant part $(4)^2 - 4(2)(-96) = 784$

$$x = \frac{-4 \pm \sqrt{784}}{4}$$

Simplify the square root. In this case 784 is a perfect square:

$$x = \frac{-4 \pm 28}{4}$$

Simplify the fraction by dividing every term by GCF, i.e., dividing –4, 28, and 4 by 4:

$x = -1 \pm 7$ which we can put together as

$x = -8 \ or \ x = 6$

The quadratic formula is the strongest tool and can be used to solve any quadratic equation. As you can see, sometimes the solutions are rational, sometimes irrational, and sometimes imaginary. **The solutions are also called the zeros or roots** of quadratic equation. **The real solutions will be shown as the x-intercepts on the graph**.

The discriminant is the part under the square root. It's called the discriminant because it can be used to tell the difference between the different types of solutions. Usually, the quadratic formula will give two solutions because of the plus/minus sign in front of the square root. If the discriminant is zero, adding or subtracting zero does not give two answers and the rest of the formula will be a fraction or rational number. If the discriminant is negative, then the square root of a negative will give an imaginary number which will be added or subtracted, giving two imaginary roots. If the discriminant it a positive number but not a perfect square, the square root of it will result in an irrational number, giving two irrational solutions. Lastly, if the discriminant it a positive number but is a perfect square, the square root of it will result in an integer, giving two rational solutions.

Quadratic Formula: $x = \dfrac{-b \pm \sqrt{b^2 - 4ac}}{2a}$ **Discriminant**

Discriminant $D = b^2 - 4ac$				
Case	Example	Example's Discriminat	Number/Type of Solutions	Graph/Number of x-intercepts
$D < 0$ Discriminant is negative	$x^2 - 2x + 3 = 0$	$D = (-2)^2 - 4 \cdot 1 \cdot 3 = -8$ $D = -8$	**2 Complex Solutions** $2\,\mathbb{C}$	zero x-ints $y = x^2 - 2x + 3$
$D = 0$ Discriminant is zero	$x^2 - 2x + 1 = 0$ If the discriminant is 0, then the quadratic can be factored as perfect square trinomial	$D = (-2)^2 - 4 \cdot 1 \cdot 1 = 0$ $D = 0$	**1 Double Rational Real Solution** $1\,\mathbb{QR}$	one x-ints $y = x^2 - 2x + 1$
$D > 0$ Discriminant is positive and **perfect square**	$x^2 - 2x - 3 = 0$ If the discriminant is positive square then the quadratic can be factored	$D = (-2)^2 - 4 \cdot 1 \cdot -3 = 16$ $D = 16$	**2 Rational Real Solutions** $2\,\mathbb{QR}$	two x-ints $y = x^2 - 2x - 3$
$D > 0$ Discriminant is positive and **NOT** perfect square	$x^2 - 2x - 1 = 0$	$D = (-2)^2 - 4 \cdot 1 \cdot -1 = 8$ $D = 8$	**2 Irrational Real Solutions** $2\,irrat.\,\mathbb{R}$	two x-ints $y = x^2 - 2x - 1$

Example 4: Find the discriminant and determine the number and type of solutions

for $3x^2 - 6x + 3 = 0$

Solution: First make sure the quadratic is in standard form set to zero. Then find the

coefficients. Here a = 3, b = –6, c = 3. Plug these values into the discriminant formula.

$D = b^2 - 4ac \quad D = (-6)^2 - 4(3)(3) = 0$

The discriminant comes out to zero, this means that there will be a zero under the square root

in the quadratic formula and the plus/minus sign in front of the square root does not give two

answers.

$D = 0\ and\ 1\,\mathbb{QR}\ sol'n\ (this\ equation\ has\ 1\ rational, real\ solution!)$

Example 5: Find the discriminant and determine the number and type of solutions

for $-2x^2 + 2x + 9 = 0$

Solution: First make sure the quadratic is in standard form set to zero. Then find the

coefficients. Here a = –2, b = 2, c = 9. Plug these values into the discriminant formula.

$D = b^2 - 4ac \quad D = (2)^2 - 4(-2)(9) = 76$

The discriminant comes out to a positive number, but it's not a square. When this is put into

the quadratic formula the square root of 76 is not rational and plus/minus sign in front of the

square root will give two answers.

$D = 76 \ and \ 2 \ irrat. \ \mathbb{R} \ sol'ns \ (this \ equation \ has \ 2 \ irrational, real \ solutions!)$

Practice Solving Quadratics by Quadratic Formula

Solve each equation with the quadratic formula.

1) $-3n^2 + 9n + 84 = 0$

$$\frac{-3}{-3} \ \frac{-3}{-3} \ \frac{-3}{-3} \ \frac{-3}{-3}$$

$n^2 - 3n - 28 = 0$

$a = 1$
$b = -3$
$c = -28$

$$\frac{-(-3) \pm \sqrt{(-3)^2 - 4(1)(-28)}}{2(1)}$$

$$\frac{3 \pm \sqrt{121}}{2} = \frac{3 \pm 11}{2}$$

$\frac{3 + 11}{2} = \boxed{7}$

$\frac{3 - 11}{2} = \boxed{-4}$

2) $-2a^2 - 6a + 36 = 0$

3) $8a^2 = -7$

4) $6n^2 - 24 = 0$

5) $3n^2 - 4n - 12 = 0$

6) $5b^2 + 10b - 22 = 0$

7) $8b^2 + 10b + 11 = 0$

8) $11x^2 - 10x + 7 = 0$

9) $-5n^2 - 2n = -23$

10) $11x^2 = -2 + 2x$

11) $-7r^2 = 11r + 9$

12) $4p^2 = 95 + p$

13) $-7k^2 + 4k + 13 = 12k - 4$

14) $-7k^2 - 14k - 4 = -4k$

15) $-x^2 + 6 = 4x + 4 - 11x^2$

16) $12r^2 - 19 = 7r^2 + 2r$

Find the discriminant of each quadratic equation then state the number and type of solutions.

17) $-5v^2 + 5v - 9 = 0$

18) $-8r^2 + r - 4 = 0$

19) $x^2 - 3x = 0$

20) $-6n^2 + n - 4 = 0$

21) $-3x^2 - 6x - 3 = 0$

22) $-8m^2 - 4m = 0$

23) $6b^2 + 3b + 6 = 0$

24) $6p^2 - 9p = 0$

25) $-9b^2 - 14b = -6b$

26) $r^2 - 3r = -4 - 7r$

27) $-6m^2 + m + 6 = -1$

28) $3n^2 - 5n = -2n^2 - 9$

From the graph of the quadratic, determine the number and type of solutions.

29)

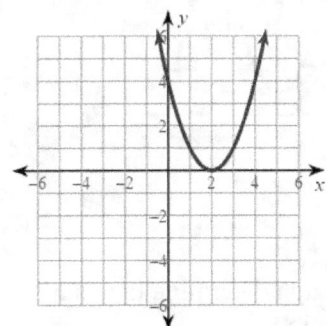

From the graph determine the number and type of solutions.
A) 2 Complex
B) 1 Rational Real
C) 2 Real

30)

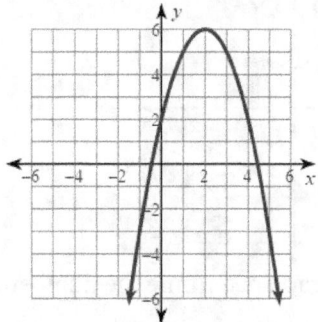

From the graph determine the number and type of solutions.
A) 2 Complex
B) 1 Rational Real
C) 2 Real

31)

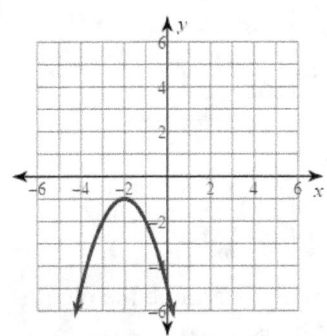

From the graph determine the number and type of solutions.
A) 2 Complex
B) 1 Rational Real
C) 2 Real

32)

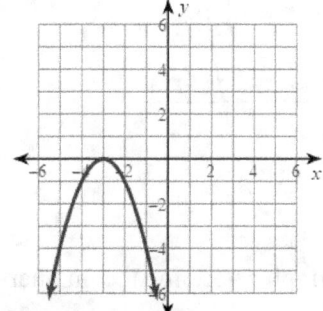

From the graph determine the number and type of solutions.
A) 2 Complex
B) 1 Rational Real
C) 2 Real

33)

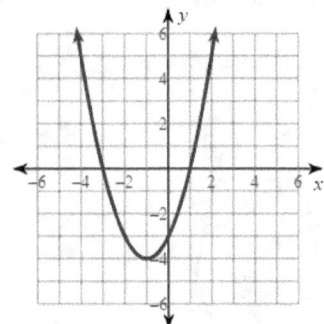

From the graph determine the number and type of solutions.
A) 2 Complex
B) 1 Rational Real
C) 2 Real

34)

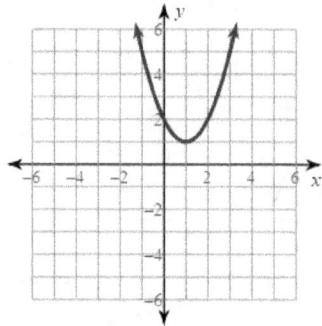

From the graph determine the number and type of solutions.
A) 2 Complex
B) 1 Rational Real
C) 2 Real

35)

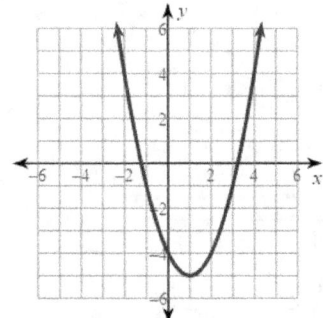

From the graph determine the number and type of solutions.
A) 2 Complex
B) 1 Rational Real
C) 2 Real

36)

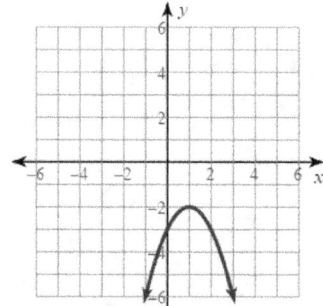

From the graph determine the number and type of solutions.
A) 2 Complex
B) 1 Rational Real
C) 2 Real

Graphing Quadratics

The last way to "solve" quadratic equations or any equation for that manner is to use a graphing calculator or software to graph and find the x-intercepts. This section will go over graphing manually while the next section will discuss using a graphing calculator or graphing software.

When graphing quadratic functions, we want the equation in standard form or vertex form. This book will focus mostly on standard form.

Standard Form

$$y = ax^2 + bx + c$$

Vertex Form

$$y = a(x - h)^2 + k$$

Graphing $y = x^2$

One way to graph any function or equation is to make a table of values, choosing values for x and substituting those values into the function to find the corresponding y values. You can pick any values for the independent variable x, but it is helpful to try to substitute zero, some negative numbers, and some positive numbers into the function.

x	y
-3	
-2	
-1	
0	
1	
2	
3	

Here's the table with our chosen x-values from -3 to 3.

$$y = (-3)^2 = 9, \, y = (-2)^2 = 4, \, y = (-1)^2 = 1, \, y = (0)^2 = 0, \, y = (1)^2 = 1, \, y = (2)^2 = 4,...$$

Filling in the table, and then plotting the (x, y) points on a coordinate plane gives:

x	y
-3	9
-2	4
-1	1
0	0
1	1
2	4
3	9

Graphing Quadratics

These are just a few points, but the actual graph of the function has an infinite number of points. All polynomials are continuous and smooth, so we can connect the points of this quadratic function with a smooth curve, ending up with graph that is a parabola:

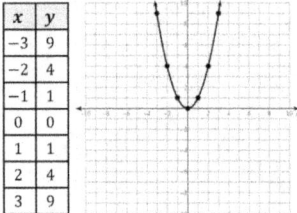

Any quadratic function will have this parabola. When the leading coefficient is positive the parabola will open up and if the leading coefficient is negative the parabola will open down. To find the y-intercepts of a function algebraically let x = 0 and solve for y. For quadratics in standard form $y = ax^2 + bx + c$ when x = 0, we get $y = a(0)^2 + b(0) + c = c$. This will be true of any polynomial in standard form, the y-intercept will be the point $(0, c)$ where c is the constant. To find the x-intercepts of a function algebraically let y = 0 and solve for x. For quadratics in standard form $y = ax^2 + bx + c$ the x-intercepts will be given by the factors. A quadratic has a line of symmetry through the vertex, because of this the vertex will always be located between the x-intercepts.

Standard Form: $y = ax^2 + bx + c$
Factored Form: $y = a(x - m)(x - n)$
Vertex Form: $y = a(x - h)^2 + k$
Opens up if $a > 0$ **Opens down if** $a < 0$

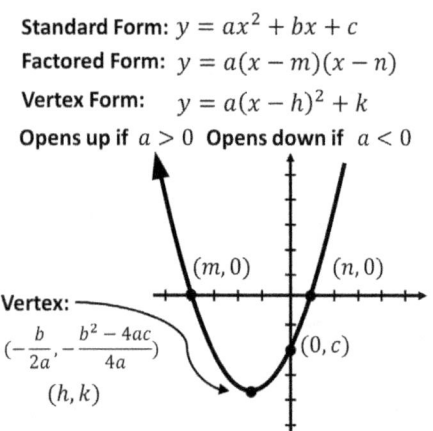

Example 1: Graph $y = -x^2 - 8x - 12$

Solution: First since this is in standard form, notice that the leading coefficient is negative, so the parabolas will open down. The constant is -12, so the y-intercept is $(0, -12)$. Try to find the x-intercepts by setting to zero and factoring:

$$0 = -x^2 - 8x - 12 = -(x^2 + 8x + 12) = -(x + 6)(x + 2) \implies x = -6, x = -2$$

So, the x-intercepts are $(-6, 0)$ and $(-2, 0)$. The vertex will be between these two when $x = -4$, plugging this into the original function:

$$y = -(-4)^2 - 8(-4) - 12 = 4 \text{ so, the vertex is at } (-4, 4)$$

Once you know what the vertex is, you can use symmetry to find other points. Since the y-intercept is at $(0, -12)$ then symmetry tells us that $(-8, -12)$ will be a point as well.

This information is enough to graph the function:

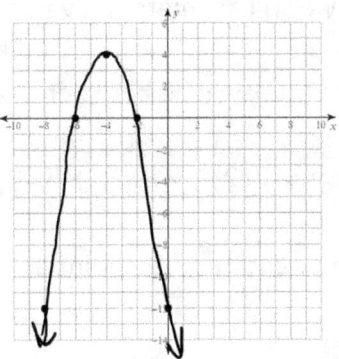

Example 2: Graph $y = 2x^2 - 8x + 5$

Solution: First since this is in standard form, notice that the leading coefficient is positive, so the parabolas will open upwards. The leading coefficient is also bigger than 1, this stretches the parabola vertically making it narrower. When the leading coefficient is between 0 and 1, the parabola widens. The constant is 5, so the y-intercept is (0,5). Try to find the x-intercepts by setting to zero and factoring:

$$0 = 2x^2 - 8x + 5$$

$2x^2 - 8x + 5$ does not factor, so we will not have rational x-intercepts. We can use the discriminant to determine if the zeros are irrational or complex. $D = (-8)^2 - 4(2)(5) = 24$. The discriminant is positive but not a perfect square, so the graph will cross the x-axis, but the roots will be irrational. Solving for the x-intercepts by the quadratic formula gives:

$$x = \frac{-(-8) \pm \sqrt{24}}{2(2)} = \frac{8 \pm 2\sqrt{6}}{4} = 2 \pm \frac{\sqrt{6}}{2}$$

The square root of 6 is between the square root of 4 and square root of 9, closer to 4. The square root of 6 will be bigger than 2, around 2.4. Dividing it by 2 gives an estimate of 1.2. So, the exact value for the x-intercepts is $(2 \pm \frac{\sqrt{6}}{2}, 0)$ but when graphing we will estimate the values to make it easier to sketch it as $(2 \pm 1.2, 0) = (0.8, 0)$ or $(3.2, 0)$. The vertex will be between the two x-intercepts with an x-value equal to $-\frac{b}{2a} = 2$. The y-value is $y = 2(2)^2 - 8(2) + 5 = -3$ The vertex is at $(2, -3)$. Using symmetry and the y-intercept, we get another free point at $(4, 5)$

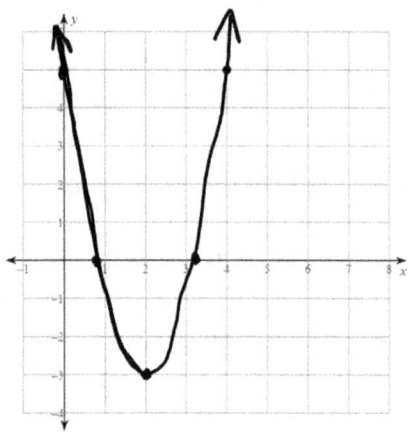

Practice Graphing Quadratics

Sketch the graph of each function.

1) $y = x^2 + 2x - 2$

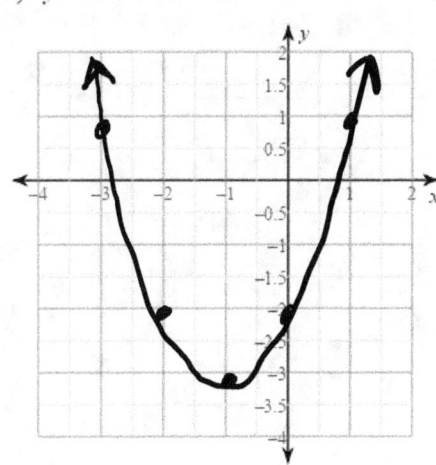

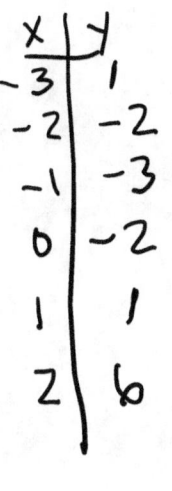

2) $y = x^2 - 6x + 5$

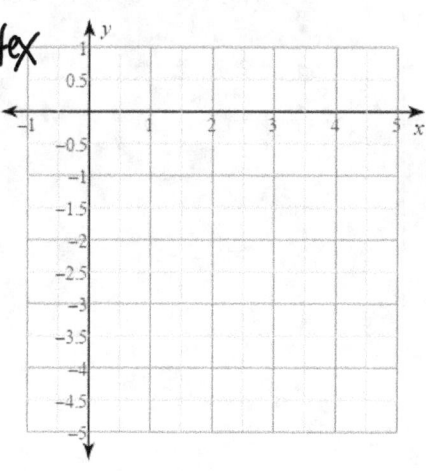

3) $y = x^2 - 8x + 13$

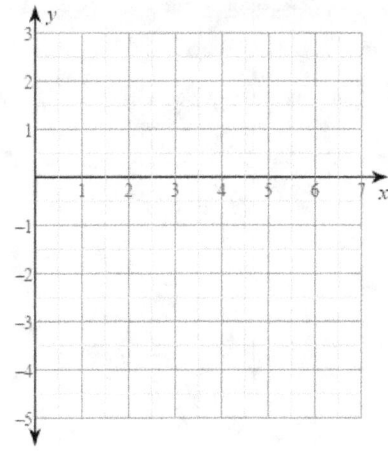

4) $y = x^2 + 4x + 7$

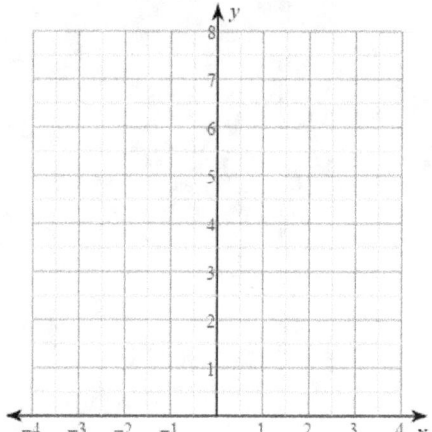

5) $f(x) = -x^2 + 8x - 18$

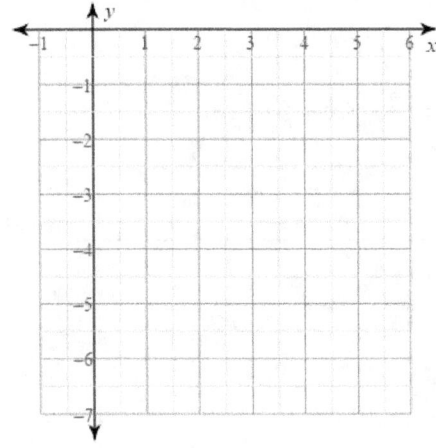

6) $f(x) = -x^2 - 4x - 2$

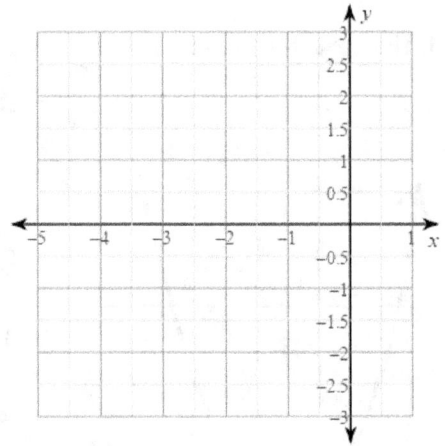

7) $f(x) = x^2 - 2x - 3$

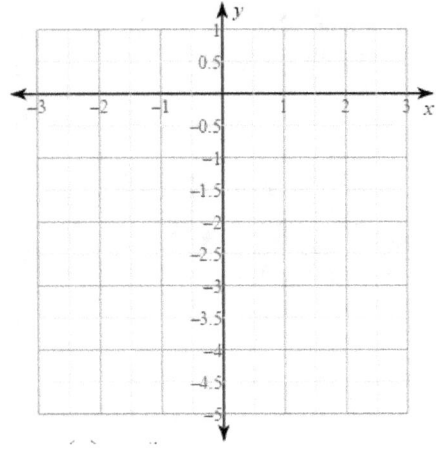

8) $f(x) = -2x^2 - 4x$

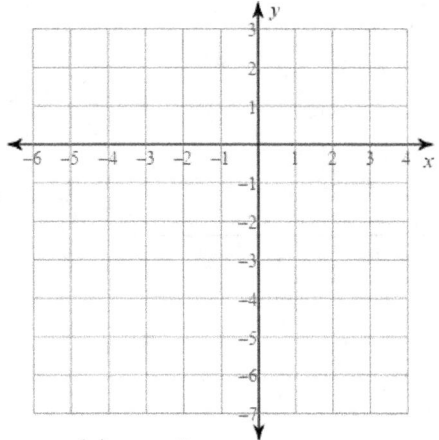

9) $f(x) = x^2 + 2x - 2$

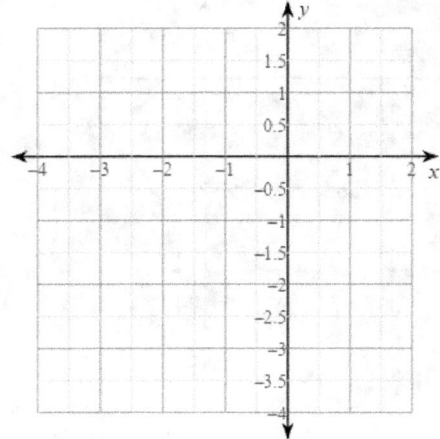

10) $f(x) = -x^2 - 2x + 1$

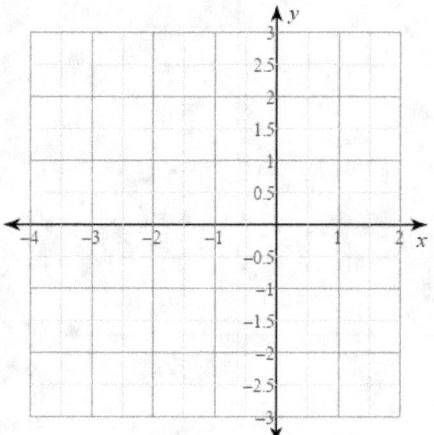

11) $f(x) = -x^2 - 2x$

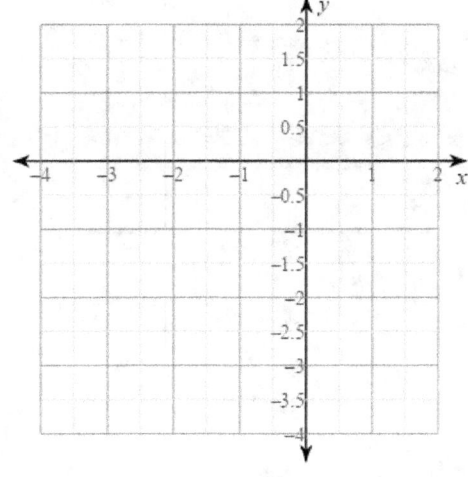

12) $f(x) = x^2 - 2x + 3$

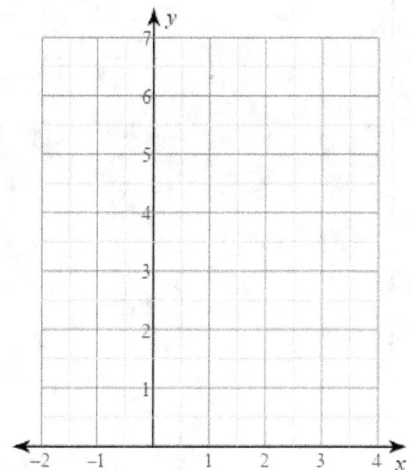

13) $f(x) = 2x^2 + 4x + 1$

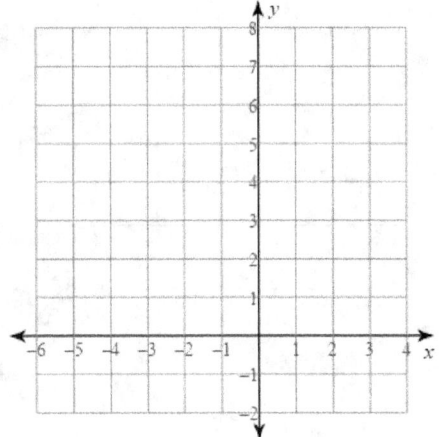

14) $f(x) = -3x^2 - 24x - 47$

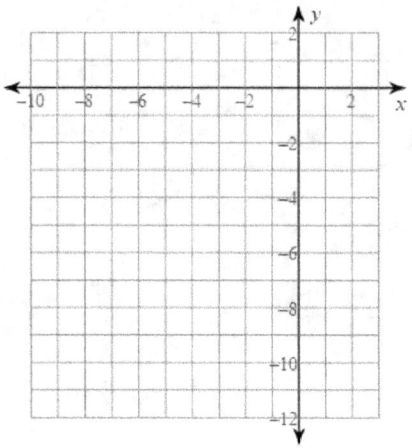

15) $f(x) = -2x^2 + 8x - 6$

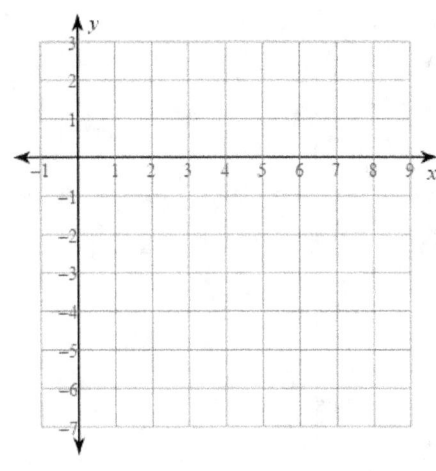

16) $f(x) = -\dfrac{1}{2}x^2 - 2x$

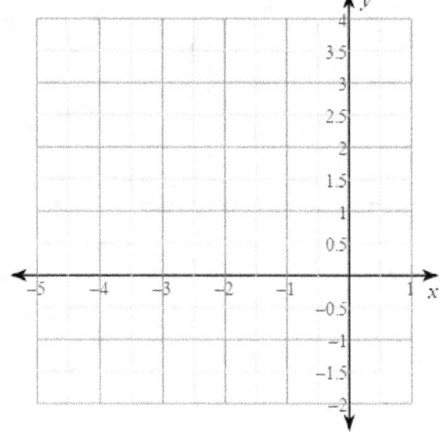

17) $y = x^2 - 2x$

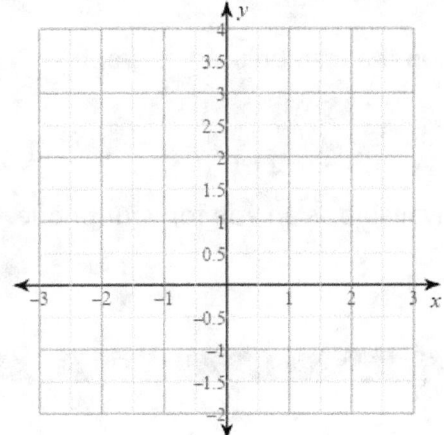

18) $y = -2x^2 + 16x - 29$

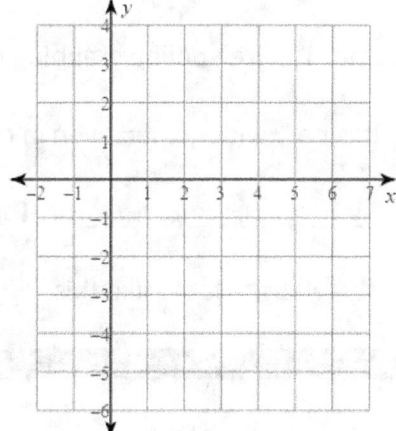

19) $y = 2x^2 - 8x + 12$

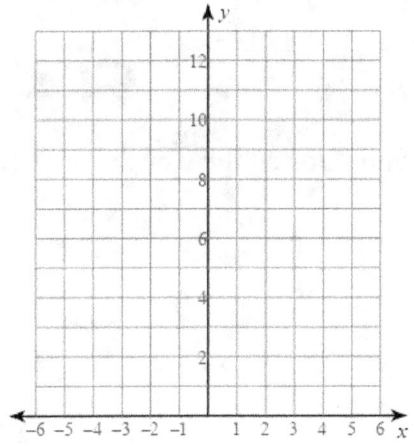

20) $y = 2x^2 - 12x + 16$

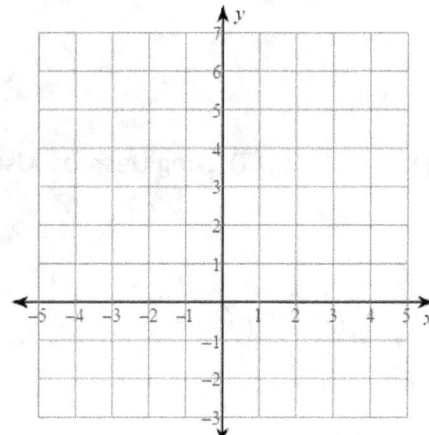

Calculator Tips Using Desmos™

Desmos is a free online graphing utility website and app that can be used to graph functions. It is now widely used even in classrooms. The website is https://www.desmos.com/calculator. The calculator is also available as an app for your phone.

The top of the website looks like this:

The bottom of the screen looks like this after pressing the keyboard tab:

Finding the GCF or LCD using Desmos. Use the function button and scroll all the way down:

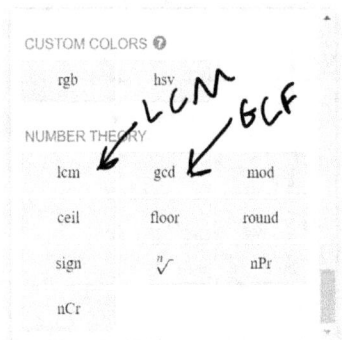

Example 1: Find the LCM and GCF of 12 and 18

$$\mathrm{lcm}(12,18)$$

$$= 36$$

$$\gcd(12,18)$$

$$= 6$$

To graph simply type in the equation on the left side. You can click on the graph and instantly find all the important points including x-intercepts, y-intercepts, and the vertices:

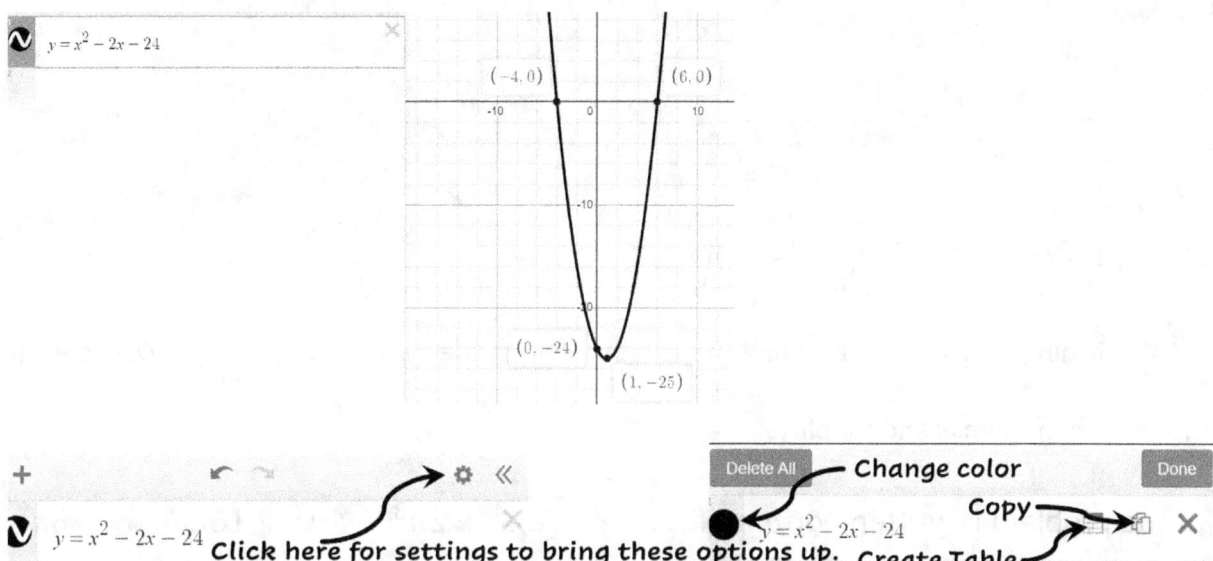

Another notable feature about Desmos is the ability to add sliders to a function. Suppose we want to graph a quadratic in vertex form $y = a(x - h)^2 + k$, we could simply type this in exactly and add sliders for the variables a, h, and k. Sliding the values for a, h, and k allows you to explore how each influences the behavior of the graph.

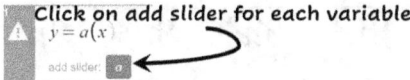

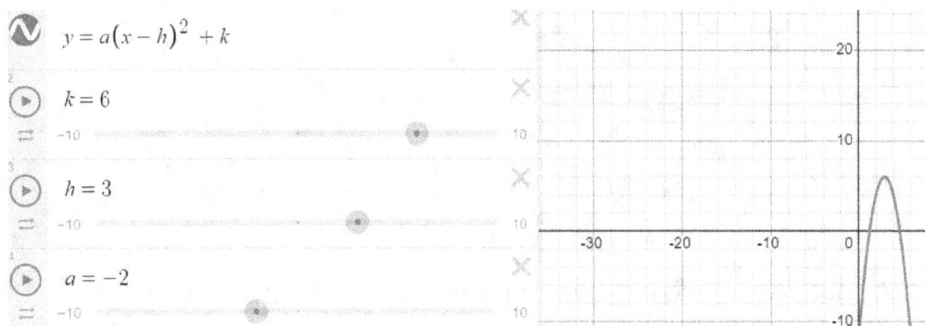

Dividing Polynomials by a Monomial

Dividing a polynomial by a monomial is done using the distributive property. Divide each term by the monomial and simplify.

If the problem is written horizontally... $(8n^4 + 2n^3 - 4n^2 + 16n - 6) \div 8n^2$

then it is usually easier to write it vertically: $\dfrac{(8n^4 + 2n^3 - 4n^2 + 16n - 6)}{8n^2}$

Divide each term by the monomial: $\dfrac{8n^4}{8n^2} + \dfrac{2n^3}{8n^2} - \dfrac{4n^2}{8n^2} + \dfrac{16n}{8n^2} - \dfrac{6}{8n^2}$

Simplify each term, reducing coefficients and subtracting exponents

$$\dfrac{8n^4}{8n^2} + \dfrac{2n^3}{8n^2} - \dfrac{4n^2}{8n^2} + \dfrac{16n}{8n^2} - \dfrac{6}{8n^2}$$

We can write our final answer using positive exponents

$$n^2 + \dfrac{n}{4} - \dfrac{1}{2} + \dfrac{2}{n} - \dfrac{3}{4n^2}$$

Alternatively, we can write our final answer using negative exponents

$$n^2 + \frac{n}{4} - \frac{1}{2} + 2n^{-1} - \frac{3}{4}n^{-2}$$

Example 1: Divide $(30x^4 - 15x^3 + 5x^2 + 50x - 4) \div 10x^2$

Solution: Divide each term by the monomial:

$$(30x^4 - 15x^3 + 5x^2 + 50x - 4) \div 10x^2 = \frac{30x^4}{10x^2} - \frac{15x^3}{10x^2} + \frac{5x^2}{10x^2} + \frac{50x}{10x^2} - \frac{4}{10x^2}$$

Simplify each fraction:

$$\frac{30x^4}{10x^2} - \frac{15x^3}{10x^2} + \frac{5x^2}{10x^2} + \frac{50x}{10x^2} - \frac{4}{10x^2} = 3x^2 - \frac{3x}{2} + \frac{1}{2} + \frac{5}{x} - \frac{2}{5x^2}$$

$$(30x^4 - 15x^3 + 5x^2 + 50x - 4) \div 10x^2 = 3x^2 - \frac{3x}{2} + \frac{1}{2} + \frac{5}{x} - \frac{2}{5x^2}$$

Practice Dividing Polynomials by a Monomial

Divide.

1) $(5x^3 + 18x^2 + 6x) \div 6x^2$

$$\frac{5x^3}{6x^2} + \frac{18x^2}{6x^2} + \frac{6x}{6x^2}$$

$$\frac{5}{6}x + 3 + \frac{1}{x}$$

2) $(10r^3 + 50r^2 + 2r) \div 10r^3$

3) $(30x^3 + 6x^2 + x) \div 6x^3$

4) $(8x^3 + 24x^2 + 16x) \div 8x^2$

5) $\left(4x^4 + x^3 + 8x^2\right) \div 4x^3$

6) $\left(2b^3 + 2b^2 + 2b\right) \div 4b^2$

7) $\left(2k^3 + 5k^2 + 8k\right) \div 8k^3$

8) $\left(3b^3 + 4b^2 + 9b\right) \div 9b^3$

9) $\left(3n^3 + 3n^2 + 9n\right) \div 9n^2$

10) $\left(8a^3 + 4a^2 + 2a\right) \div 4a$

11) $\left(5x^4 + 2x^3 + 8x^2\right) \div 8x^3$

12) $\left(9a^3 + 9a^2 + 18a\right) \div 9a$

13) $\left(8x^3 + 12x^2 + 4x\right) \div 4x^2$

14) $\left(24m^3 + 24m^2 + 2m\right) \div 6m^2$

15) $\left(12p^4 + 6p^3 + 4p^2\right) \div 6p$

16) $\left(5a^6 + 5a^5 + 50a^4\right) \div 10a^3$

17) $\left(5p^3 + 2p^2 + 16p\right) \div 8p^2$

18) $\left(5m^3 + m^2 + 5m\right) \div 10m^2$

19) $\left(30r^4 + 2r^3 + 2r^2\right) \div 6r$

20) $\left(8n^4 + 5n^3 + n^2\right) \div 4n$

Dividing Polynomials by Long Division

To understand how to divide any polynomial by any other polynomial it is important to understand the arithmetic version of long division.

$$\text{DIVIDEND} \div \text{DIVISOR} = \text{QUOTIENT} \qquad \text{DIVISOR} \overline{\smash{)}\text{DIVIDEND}}^{\text{QUOTIENT}}$$

Dividing Polynomials by Long Division

Example 1: Divide $145024 \div 12$ using long division

Solution: First set up the division box the dividend (number to be divided) goes inside the box and the divisor (number we are dividing by) goes to left.

$$12\overline{)145024}$$

Now we think how many times does 12 go into 1? It doesn't, so we divide 12 into 14 which goes in 1 time, and we write 1 at the top:

$$\begin{array}{r} 1 \\ 12\overline{)145024} \end{array}$$

Multiply that 1 by the divisor 12 which gives 12 and place it below the dividend.

$$\begin{array}{r} 1 \\ 12\overline{)145024} \\ \underline{12} \end{array}$$

Subtract from the dividend:

$$\begin{array}{r} 1 \\ 12\overline{)145024} \\ \underline{12} \\ 2 \end{array}$$

Bring the next number down:

$$\begin{array}{r} 1 \\ 12\overline{)145024} \\ \underline{12} \\ 25 \end{array}$$

How many times does 12 go into 25? 2 times. Place 2 next in the quotient:

$$
\begin{array}{r}
12 \\
12 \overline{\smash{)}145024} \\
\underline{12} \\
25 \\
\underline{24} \\
1
\end{array}
$$

Multiply 2 by 12 and place below, subtract, bring down next number:

Keep repeating this process...

$$
\begin{array}{r}
12085 \; R4 \\
12 \overline{\smash{)}145024} \\
\underline{12} \\
25 \\
\underline{24} \\
10 \\
\underline{0} \\
102 \\
\underline{96} \\
64 \\
\underline{60} \\
4
\end{array}
$$

Eventually reaching the end until all digits in the dividend are brought down. The last number is a remainder which can be written as a remainder or fraction. To get a fraction, take the remainder divided by the divisor. (If you wanted a decimal answer, you would add a decimal point and zeros in the dividend, repeating the process until you get a reminder of zero or a pattern repeats.)

Now we apply a similar algorithm for long division of polynomials. We can divide a polynomial by another polynomial, usually a binomial. When dividing polynomials, the polynomial should be in standard form with descending exponents.

Dividing Polynomials by Long Division

Example 2: Divide $(-87x^2 + 10x^3 + 46 - 31x) \div (x - 9)$

Solution: Put the dividend into standard form with descending exponents:

$$(-87x^2 + 10x^3 + 46 - 31x) \div (x - 9) = (10x^3 - 87x^2 - 31x + 46) \div (x - 9)$$

Set up the division box:

$$x - 9 \overline{\smash{)}10x^3 - 87x^2 - 31x + 46}$$ Divide the first term of the dividend by the first term of

the divisor. $10x^3$ divided by x is $10x^2$.

$$\begin{array}{r} 10x^2 \\ x - 9 \overline{\smash{)}10x^3 - 87x^2 - 31x + 46} \end{array}$$ Multiply $10x^2$ by the divisor and place below the dividend.

$$\begin{array}{r} 10x^2 \\ x - 9 \overline{\smash{)}10x^3 - 87x^2 - 31x + 46} \\ 10x^3 - 90x^2 \end{array}$$ Change the signs and combine like terms:

$$\begin{array}{r} 10x^2 \\ x - 9 \overline{\smash{)}10x^3 - 87x^2 - 31x + 46} \\ -10x^3 + 90x^2 \end{array}$$ Bring the next term down.

$$\begin{array}{r} 10x^2 \\ x - 9 \overline{\smash{)}10x^3 - 87x^2 - 31x + 46} \\ -10x^3 + 90x^2 \\ 3x^2 - 31x \end{array}$$ Divide the first term of the bottom by the first term of the

divisor. $3x^2$ divided by x is $3x$.

$$\begin{array}{r} 10x^2 + 3x \\ x - 9 \overline{\smash{)}10x^3 - 87x^2 - 31x + 46} \\ -10x^3 + 90x^2 \\ 3x^2 - 31x \\ 3x^2 - 27x \end{array}$$ Subtract, combine like terms, bring the next term down:

$$
\begin{array}{r}
10x^2 + 3x \\
x - 9 \overline{\smash{\big)}\, 10x^3 - 87x^2 - 31x + 46} \\
\underline{-10x^3 + 90x^2} \\
3x^2 - 31x \\
\underline{-3x^2 + 27x} \\
-4x + 46
\end{array}
$$

Divide the first term of the bottom by the first term of the

divisor. *–4x* divided by *x* is *–4*.

$$
\begin{array}{r}
10x^2 + 3x - 4 \\
x - 9 \overline{\smash{\big)}\, 10x^3 - 87x^2 - 31x + 46} \\
\underline{-10x^3 + 90x^2} \\
3x^2 - 31x \\
\underline{-3x^2 + 27x} \\
-4x + 46 \\
\underline{-4x + 36}
\end{array}
$$

Subtract, combine like terms, since there is no term to

bring down the difference is the remainder:

$$
\begin{array}{r}
10x^2 + 3x - 4 \text{ R }10 \\
x - 9 \overline{\smash{\big)}\, 10x^3 - 87x^2 - 31x + 46} \\
\underline{-10x^3 + 90x^2} \\
3x^2 - 31x \\
\underline{-3x^2 + 27x} \\
-4x + 46 \\
\underline{+4x - 36} \\
10
\end{array}
$$

$$(10x^3 - 87x^2 - 31x + 46) \div (x - 9) = 10x^2 + 3x - 4 \; R10 \; or \; 10x^2 + 3x - 4 + \frac{10}{x - 9}$$

If the remainder comes out to zero, then the divisor is a **factor** of the dividend.

Example 3: Determine if $(x + 2)$ is a factor of $(6x^3 + 9x^2 + 3x + 18)$

Solution: Divide $(6x^3 + 9x^2 + 3x + 18)$ by $(x + 2)$. If the remainder is 0 then $(x + 2)$ is a

factor.

$$
\begin{array}{r}
6x^2 - 3x + 9 \ \text{R} \ 0 \\
x + 2 \ \overline{\big)\ 6x^3 + 9x^2 + 3x + 18} \\
-6x^3 + 12x^2 \\
\hline
-3x^2 + 3x \\
+3x^2 + 6x \\
\hline
9x + 18 \\
-9x + 18 \\
\hline
0
\end{array}
$$

Yes, $(x + 2)$ is a factor of $(6x^3 + 9x^2 + 3x + 18)$

Example 4: Divide $(10x^5 - 70x^4 - 2x + 11) \div (x - 7)$

Solution: Notice that some terms in the dividend are missing, just put zeros in for the missing

terms.

$$
\begin{array}{r}
10x^4 \qquad\qquad\qquad\qquad\quad -2 \ \ R \ -3 \\
x - 7 \ \overline{\big)\ 10x^5 - 70x^4 + 0x^3 + 0x^2 - 2x + 11} \\
10x^5 - 70x^4 \\
\hline
0x^4 + 0x^3 \\
0x^4 + 0x^3 \\
\hline
0x^3 + 0x^2 \\
0x^3 + 0x^2 \\
\hline
-2x + 11 \\
-2x + 14 \\
\hline
-3
\end{array}
$$

$$(10x^5 - 70x^4 - 2x + 11) \div (x - 7) = 10x^4 - 2 - \frac{3}{x - 7}$$

Practice Dividing Polynomials by Long Division

Divide.

1) $(3n^3 + 36n^2 + 51n - 8) \div (3n + 6)$

$$
\begin{array}{r}
n^2 + 10n - 3 \ R10 \\
3n+6 \overline{\big)\ 3n^3 + 36n^2 + 51n - 8} \\
-\ 3n^3 + 6n^2 \\
\hline
30n^2 + 51n \\
30n^2 - 60n \\
\hline
-9n - 8 \\
+ 9n + 18 \\
\hline
10
\end{array}
$$

$n^2 + 10n - 3 + \dfrac{10}{3n+6}$

2) $(9k^3 - 98k^2 + 62k + 19) \div (9k - 8)$

3) $(35x^3 + 23x^2 - 22x + 83) \div (5x + 9)$

4) $(16x^3 - 56x^2 + 56x - 47) \div (4x - 10)$

5) $(8r^4 - 46r^3 - 68r^2 + 2r - 3) \div (8r + 2)$

6) $(7m^4 - 50m^3 + 27m^2 + 66m - 42) \div (7m - 8)$

7) $(4a^4 + 38a^3 + 44a^2 + 30a + 49) \div (4a + 6)$

8) $(8x^4 - 2x^3 - 14x^2 - 46x - 38) \div (8x + 6)$

9) $(3x^5 - 5x^4 + 2x^3 + 58x^2 - 52x - 57) \div (3x + 7)$

10) $(6n^5 - 45n^4 - 57n^3 + 30n^2 + 21n + 53) \div (6n + 9)$

11) $\left(3m^5 + 7m^4 + 7m^3 - 11m^2 + 7m + 31\right) \div \left(3m + 4\right)$

12) $\left(8n^5 + 9n^4 - 39n^3 - 13n^2 + 55n + 5\right) \div \left(8n + 1\right)$

13) $\left(100x^3 - 100x^2 - 10x\right) \div \left(10x - 10\right)$

14) $\left(72a^3 - 39a^2 + 66a + 11\right) \div \left(8a + 1\right)$

15) $\left(63m^2 + 142m + 78\right) \div \left(7m + 8\right)$

16) $\left(3b^4 + 15b^3 + 6b^2 - 12b + 77\right) \div \left(3b + 9\right)$

17) $\left(4x^5 - 6x^4 - 10x^3 - 4\right) \div \left(4x - 10\right)$

18) $\left(8x^4 + 76x^3 + 8x^2 - 72x + 28\right) \div \left(8x - 4\right)$

19) $\left(16n^2 + 5n^3 + 26 + 37n\right) \div \left(5n + 6\right)$

20) $\left(-42p + 26p^2 + 4 + 10p^3\right) \div \left(-4 + 10p\right)$

State if the given binomial is a factor of the given polynomial.

21) $(6n^4 + 4n^3 + 9) \div (6n + 4)$

22) $(4x^3 - 5x^2 - 10x + 8) \div (x - 2)$

23) $(21p^2 - 12 + 6p^3 - 15p) \div (3 + 6p)$

24) $(-63 + x^2 + 71x + 10x^3) \div (-9 + 10x)$

25) $(5x^2 + 52x + 18) \div (5x + 2)$

26) $(15k^2 + 22k - 5) \div (3k + 5)$

27) $(8p^2 + 61p + 35) \div (8p + 5)$

28) $(8x^3 - 38x^2 - 10 - 50x) \div (2 + 8x)$

29) $(2v^4 + 5v^3 - 12v^2 + 17v - 45) \div (2v + 9)$

30) $(9r^4 - 18r^3 - 9) \div (3r - 6)$

31) $(-20 - 58m + 63m^2) \div (9m + 2)$

32) $(8r^5 + 2r^4 - 48r^2 - 12r + 6) \div (8r + 2)$

Dividing Polynomials by Synthetic Division

Synthetic division is an easier way to divide polynomials, but it can only be used when the divisor is of the form *(x – m)*. This is done by putting *m* in a box in the top left corner, then writing down the coefficients and constant of the dividend, leaving some space underneath, then drawing a line where the answer will go below it. By switching the sign and changing *(x – m)* into *m* or changing *(x + m)* into *–m*, we avoid changing the signs for the rest of the problem like we do for normal long division.

Example 1: Divide $(x^4 + 9x^3 + 10x^2 - 16x - 15) \div (x + 2)$ by synthetic division

Solution:

$$\begin{array}{c|ccccc} -2 & 1 & 9 & 10 & -16 & -15 \\ \hline & & & & & \end{array}$$

Now that it is set up, we begin by bringing the first number down:

$$\begin{array}{c|ccccc} -2 & 1 & 9 & 10 & -16 & -15 \\ \hline & 1 & & & & \end{array}$$

Multiply this number by the number in the box and write it below the next number.

$$\begin{array}{c|ccccc} -2 & 1 & 9 & 10 & -16 & -15 \\ & & -2 & & & \\ \hline & 1 & & & & \end{array}$$

Multiply Going Up

Add the numbers going down:

$$\begin{array}{c|ccccc} -2 & 1 & 9 & 10 & -16 & -15 \\ & & -2 & & & \\ \hline & 1 & 7 & & & \end{array}$$

Add Going Down

Repeat the process, multiply this number by the number in the box and write it below the next number.

$$
\begin{array}{r|rrrrr}
-2 & 1 & 9 & 10 & -16 & -15 \\
& & -2 & -14 & & \\
\hline
& 1 & 7 & & \text{\textbf{Multiply Going Up}}
\end{array}
$$

Add the numbers going down:

$$
\begin{array}{r|rrrrr}
-2 & 1 & 9 & 10 & -16 & -15 \\
& & -2 & -14 & & \\
\hline
& 1 & 7 & -4 & \text{\textbf{Add Going Down}}
\end{array}
$$

Keep repeating, multiplying to go up, adding to go down until you get to the end. The last number on the right is the remainder, the other numbers on the bottom are the coefficients and constant of the answer. The degree of the answer is one less than the degree of the starting polynomial (dividend). Since we started with a degree 4 polynomial and divided by a linear degree 1, the answer will be a cubic polynomial of degree 3:

$$
\begin{array}{r|rrrrr}
-2 & 1 & 9 & 10 & -16 & -15 \\
& & -2 & -14 & 8 & 16 \\
\hline
& 1 & 7 & -4 & -8 & \boxed{1} \quad \longleftarrow \text{\textbf{This number is the remainder}}
\end{array}
$$

These numbers are the coefficients

$$(x^4 + 9x^3 + 10x^2 - 16x - 15) \div (x + 2) = x^3 + 7x^2 - 4x - 8 + \dfrac{1}{x + 2}$$

Example 2: Divide $(9x^4 - 49x^3 + 20x^2 + 2) \div (x - 5)$ by synthetic division

Solution: This has a missing x term, so a zero will go in that spot. Since the divisor is *(x – 5), 5* will go in the divisor box:

$$
\begin{array}{r|rrrrr}
5 & 9 & -49 & 20 & 0 & 2 \\
  &   & 45 & -20 & 0 & 0 \\
\hline
  & 9 & -4 & 0 & 0 & 2
\end{array}
$$

This number is the remainder

These numbers are the coefficients

Synthetic division can only be used if the divisor is a binomial of the form *(x – m)* but if the divisor is a binomial of the form *(px – m)*. To do this we divide both the divisor and dividend by p, then the divisor becomes $(x - \frac{m}{p})$ and we can set up the synthetic division.

Example 3: Divide $(6x^3 - 8x^2 - 40x + 10) \div (6x - 2)$ by synthetic division

Solution: We rewrite the problem by dividing everything by 6:

$$(x^3 - \frac{4}{3}x^2 - \frac{20}{3}x + \frac{5}{3}) \div (x - \frac{1}{3})$$

$$
\begin{array}{r|rrrr}
\frac{1}{3} & 1 & -\frac{4}{3} & -\frac{20}{3} & \frac{5}{3} \\
  &   & \frac{1}{3} & -\frac{1}{3} & -\frac{7}{3} \\
\hline
  & 1 & -1 & -7 & -\frac{2}{3}
\end{array}
$$

The coefficients come out nice, but the remainder is a fraction. This remainder will be written over the new divisor, multiplying the top and bottom by the LCD of 3 it should be simplified as:

$$\frac{-\frac{2}{3}}{x - \frac{1}{3}} = \frac{-2}{3x - 1}$$

So, the final answer to our problem becomes:

$$(6x^3 - 8x^2 - 40x + 10) \div (6x - 2) = x^2 - x - 7 - \frac{2}{3x - 1}$$

Practice Dividing Polynomials by Synthetic Division

Divide by synthetic division

1) $(x^4 + 5x^3 - 6x^2 - 15x - 13) \div (x + 1)$

$$
\begin{array}{c|ccccc}
-1 & 1 & 5 & -6 & -15 & -13 \\
 & & -1 & -4 & 10 & 5 \\
\hline
 & 1 & 4 & -10 & -5 & \boxed{-8}
\end{array}
$$

$x^3 + 4x^2 - 10x - 5 - \dfrac{8}{x+1}$

2) $(4x^3 - 31x^2 + 14x + 46) \div (x - 7)$

3) $(7m^4 - 15m^3 - 11m^2 - 23m + 14) \div (m - 3)$

4) $(a^3 - 2a^2 + 11a - 7) \div (a - 1)$

5) $(x^3 + x^2 - 16x - 22) \div (x - 4)$

6) $(k^5 - 4k^4 - 12k^3 - 70k^2 + 56k - 39) \div (k - 7)$

7) $(n^4 - 7n^3 + 4n^2 + 14n + 5) \div (n - 2)$

8) $(n^3 + 2n^2 - 43n + 32) \div (n + 8)$

9) $(v^4 - 9v^3 + 7v^2 + 41v + 55) \div (v - 7)$

10) $(r^4 - r^3 - 25r^2 - 27r - 33) \div (r + 4)$

11) $(5a^4 + 52a^3 + 64a^2 + 2a - 62) \div (a + 9)$

12) $(9p^3 + 45p^2 + 1) \div (p + 5)$

13) $\left(6a^4 + 22a^3 + 13a^2 - 20a - 17\right) \div \left(a + 2\right)$

14) $\left(r^4 - 4r^3 + 9r - 27\right) \div \left(r - 4\right)$

15) $\left(x^4 - 70x^2 + 50x - 24\right) \div \left(x - 8\right)$

16) $\left(r^3 - 4r^2 - 56r + 26\right) \div \left(r + 6\right)$

17) $\left(18v^4 + 54v^3 + 82v^2 + 18v - 14\right) \div \left(6v + 4\right)$

18) $\left(4k^4 - 9k^3 - 2k^2 - 31k + 3\right) \div \left(4k - 1\right)$

19) $\left(12x^4 - 68x^3 + 84x^2 + 4x - 85\right) \div \left(2x - 8\right)$

20) $\left(28n^4 + 10n^3 + 16n^2 + 59n + 27\right) \div \left(7n + 6\right)$

21) $\left(9v^5 + v^4 + 81v^3 - 46v - 4\right) \div \left(9v + 1\right)$

22) $\left(6x^3 + 4x^2 + 7\right) \div \left(6x + 4\right)$

23) $\left(4a^5 - 6a^4 + 20a^2 - 62a + 39\right) \div \left(4a - 6\right)$

24) $\left(2n^3 - 13n^2 - 51n - 9\right) \div \left(2n + 5\right)$

Graphs of Polynomials

Linear and quadratic functions are part of the family called polynomial functions.

Polynomial functions can be written in the form:

$$P(x) = a_n x^n + a_{n-1} x^{n-1} + \cdots + a_2 x^2 + a_1 x + a_0$$

The **degree** of a polynomial is the highest exponent n. The zeros are the solutions to P(x)=0. The

real zeros are the same as the x-intercepts. The **extrema** are the high points and low points,

peaks and valleys (sometimes these are called turning points). All polynomial functions have

graphs that are continuous, smooth, and stretch all the way left and right. Continuous means

the graph does not have any holes, gaps, or breaks. It can be drawn without lifting your

pen/pencil. The graph is smooth in that it has no corners, cusps, jagged or rough edges or

vertical tangents (in calculus this is called differentiable). The domain of all polynomial

functions is negative infinity to positive infinity. This means you could plug any real number in

for the input x. The range, or y values, for all odd degree polynomials is also all reals.

Name	Equation	Deg	Graph a > 0	Graph a < 0	Max # of Zeros	Max # of Extrema
Constant	$y = a$	0			0	0
Linear	$y = ax + b$	1			1	0
Quadratic	$y = ax^2 + bx + c$	2			2	1
Cubic	$y = ax^3 + \cdots$	3			3	2
Quartic	$y = ax^4 + \cdots$	4			4	3
Quintic	$y = ax^4 + \cdots$	5			5	4

Practice Graphs of Polynomials

1) Identify the degree of the polynomial, then name and whether the leading coefficient is positive or negative.

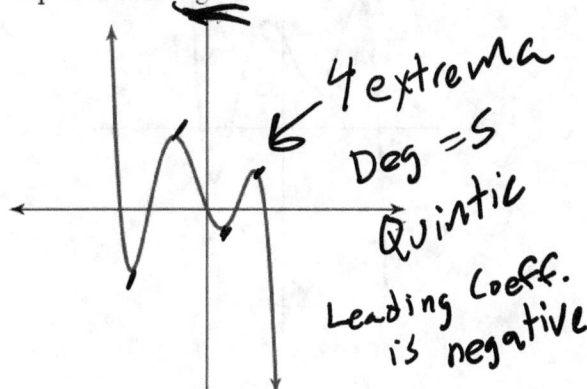

4 extrema

Deg = 5

Quintic

Leading Coeff. is negative

2) Identify the degree of the polynomial, then name and whether the leading coefficient is positive or negative.

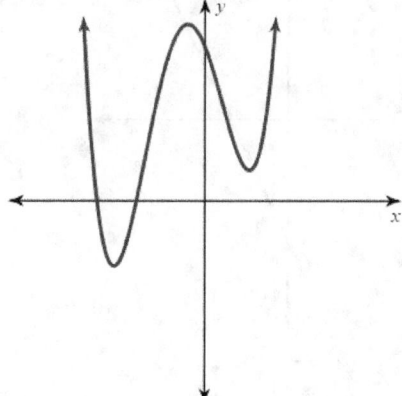

3) Identify the degree of the polynomial, then name and whether the leading coefficient is positive or negative.

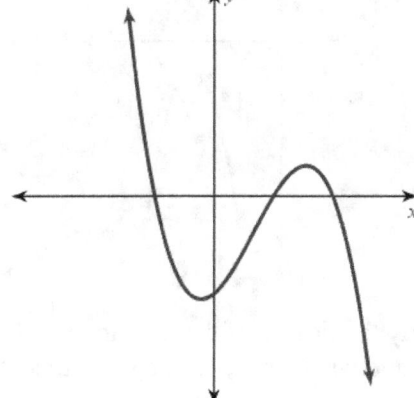

4) Identify the degree of the polynomial, then name and whether the leading coefficient is positive or negative.

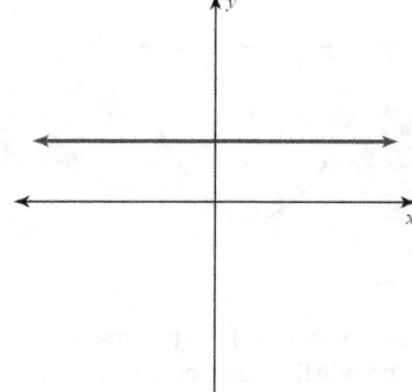

5) Identify the degree of the polynomial, then name and whether the leading coefficient is positive or negative.

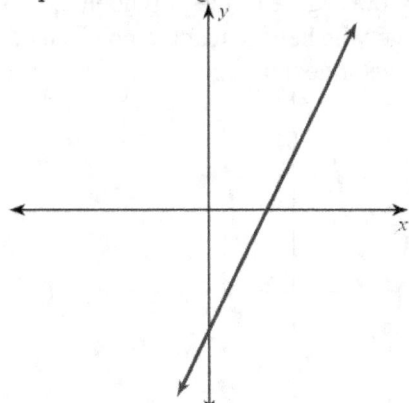

6) Identify the degree of the polynomial, then name and whether the leading coefficient is positive or negative.

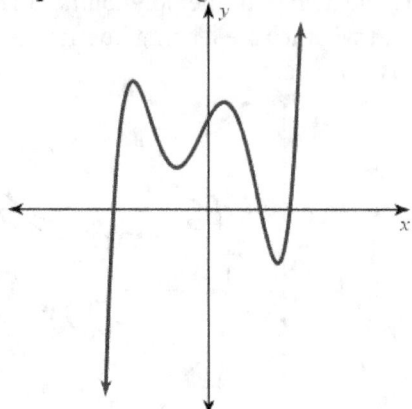

7) Identify the degree of the polynomial, then name and whether the leading coefficient is positive or negative.

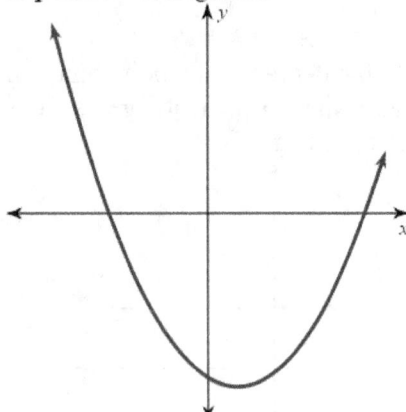

8) Identify the degree of the polynomial, then name and whether the leading coefficient is positive or negative.

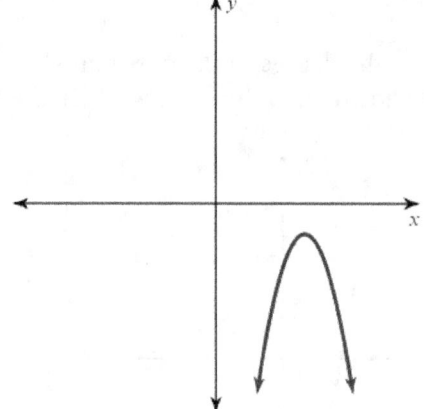

9) Identify the degree of the polynomial, then name and whether the leading coefficient is positive or negative.

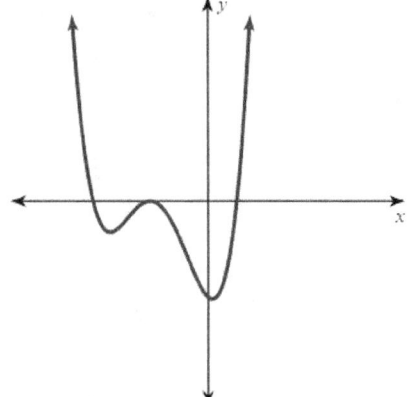

10) Identify the degree of the polynomial, then name and whether the leading coefficient is positive or negative.

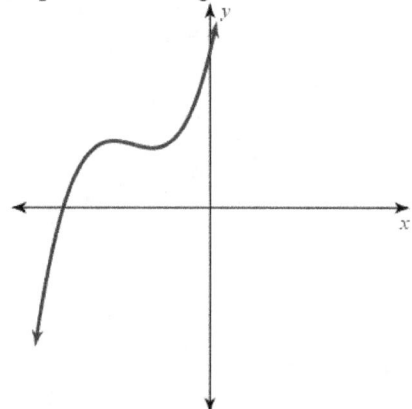

11) Identify the degree of the polynomial, then name and whether the leading coefficient is positive or negative.

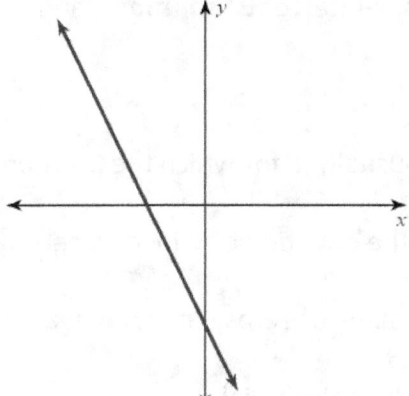

12) Identify the degree of the polynomial, then name and whether the leading coefficient is positive or negative.

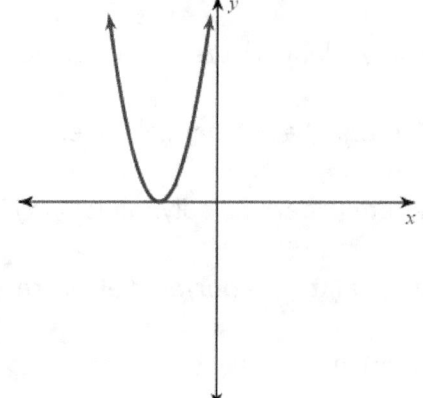

13) Identify the degree of the polynomial, then name and whether the leading coefficient is positive or negative.

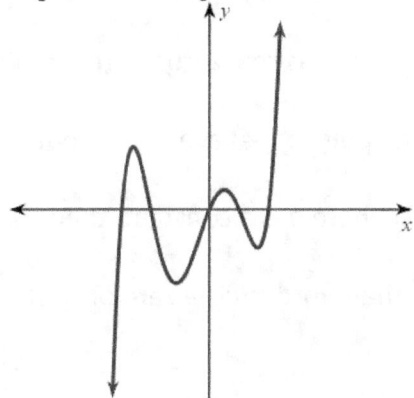

14) Identify the degree of the polynomial, then name and whether the leading coefficient is positive or negative.

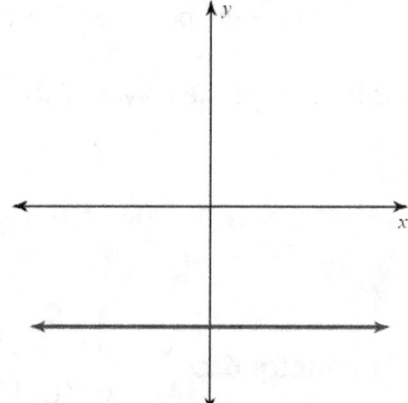

15) Identify the degree of the polynomial, then name and whether the leading coefficient is positive or negative.

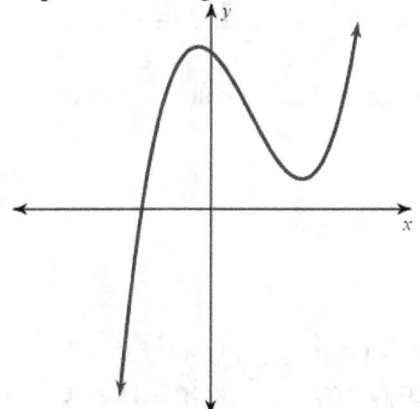

16) Identify the degree of the polynomial, then name and whether the leading coefficient is positive or negative.

Zeros, Roots, and Factors

In the context of polynomials, the terms **zeros**, **roots**, **x-intercepts**, **solutions,** and **factors** are all related and interconnected.

Zeros of a polynomial are the values of the variable, usually x, for which the polynomial evaluates to zero. In other words, they are the **solutions** to the equation obtained by setting the polynomial equal to zero. **Roots** of a polynomial are the same as zeros. They represent the values of x that satisfy the equation $P(x) = 0$, where $P(x)$ is the polynomial.

The x-intercepts of a polynomial are the points on the x-axis where the graph of the polynomial intersects or crosses the x-axis. Geometrically, these are the points $(x, 0)$ where the polynomial evaluates to zero. X-intercepts are essentially the same as real zeros or real roots.

Factors of a polynomial are expressions that divide the polynomial evenly without leaving a remainder. If a polynomial $P(x)$ has a factor $(x - a)$, where a is a constant, then substituting a into the polynomial will result in $P(a) = 0$. In other words, a is a **zero** or **root** of the polynomial.

The relationship between zeros, roots, x-intercepts, and factors is summarized as follows: **Zeros/Roots/Solutions** correspond to the values of x where the polynomial equals zero, **x-intercepts** are the points on the x-axis where the graph crosses it, and factors are expressions that divide the polynomial evenly, leading to the zeros/roots/solutions of the polynomial.

$$(x - a) \text{ is a } \textbf{factor} \iff \begin{array}{c} a \text{ is a } \textbf{real solution} \\ \text{to } P(x) = 0 \end{array} \iff \begin{array}{c} a \text{ is a } \textbf{real} \\ \textbf{root/zero} \end{array} \iff \begin{array}{c} (a, 0) \text{ is an} \\ \textbf{x-intercept} \end{array}$$

You can find the zeros/roots of a polynomial by factoring it and setting each factor to zero,

solving for x.

Example 1: Find the zeros of the polynomial $4x(x-2)(x+3)(2x-3)$

Solution: Since this is already factored, we set each factor to zero.

The factors are: 4, x, x − 2, x + 3, and 2x − 3.

4 is a constant and never equal to zero, so you don't get any zeros from constant factors, only

factors that have a variable. Setting the remaining factors to zero:

x = 0 or x − 2 = 0 or x + 3 = 0 or 2x − 3 = 0. The solutions are:

x = 0, x = 2, x = − 3, x=3/2

Example 2: Find the solutions to $x^2 - x - 12 = 0$

Solution: When a polynomial is set to zero, the solutions to the equation are the same as the

zeros/roots and the same as the x-intercepts when we start graphing. We can find the zeros by

factoring the quadratic polynomial.

$x^2 - x - 12 = (x-4)(x+3)$ after factoring, set the factors to zero and solve:

$x - 4 = 0 \ or \ x + 3 = 0$ which gives solutions:

$x = 4 \ or \ x = -3$

Example 3:

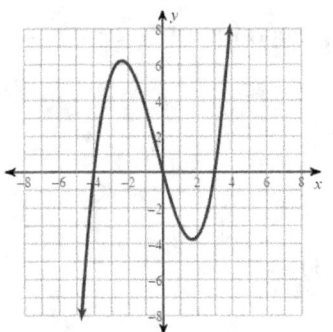

f

Identify the zeros:

What are the factors:

Solution: Look at the graph and determine where the x-intercepts (where it crosses or touches

the x-axis) are: (– 4, 0), (0, 0), (3, 0)

The zeros are the same as the x-values of the x-intercepts: zeros = – 4, 0, 3

 The factors: (x + 4), x, and (x – 3).

Example 4:

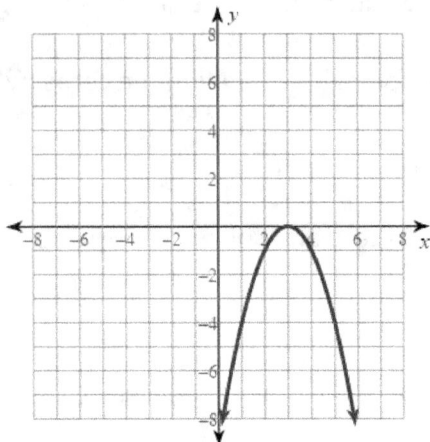

Identify the zeros:

What are the factors:

Solution: This graph is a quadratic (degree 2), but it only has one x-intercept at (3, 0)

There is one zero at the same x-value of the x-intercept: zeros = 3

This gives us only one factor (x – 3) but a quadratic cannot be formed from only one linear factor, it needs two. So, this factor has to be squared giving us a double root at 3. Since it is open down, the leading coefficient needs to be negative. The factors are: $-1(x - 3)^2$.

Practice Zeros, Roots, and Factors

1) A polynomial is factored as:
   $2x(x - 4)(x + 5)(3x - 4)$
   Find all the zeros

   $X = 0, X - 4 = 0, X + 5 = 0, 3X - 4 = 0$

   $X = 0, 4, -5, \frac{4}{3}$

2) A polynomial is factored as:
   $2(x + 4)(x - 5)(4x + 3)$
   Find all the zeros

3) A polynomial is factored as:
   $4(x - 3)(x + 5)(2x + 7)$
   Find all the x-intercepts

4) A polynomial is factored as:
   $-4x(x + 3)(x - 2)(2x - 1)$
   Find all the x-intercepts

5) A polynomial has zeros at:
   $-7, 0, 2, -\frac{6}{5}$
   Find all the factors

6) A polynomial has real roots at:
   $-2, -\frac{3}{2}, \frac{2}{5}, 6$
   Find all the factors

7)

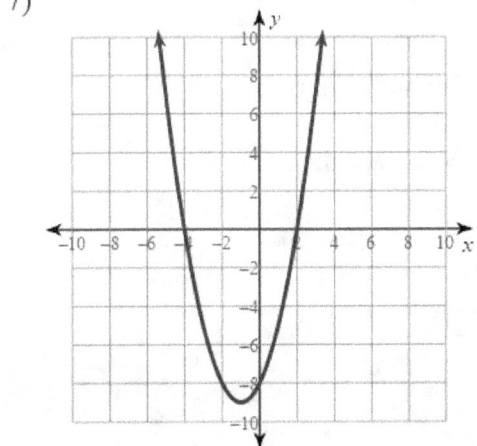

Identify the zeros:

What are the factors:

8)

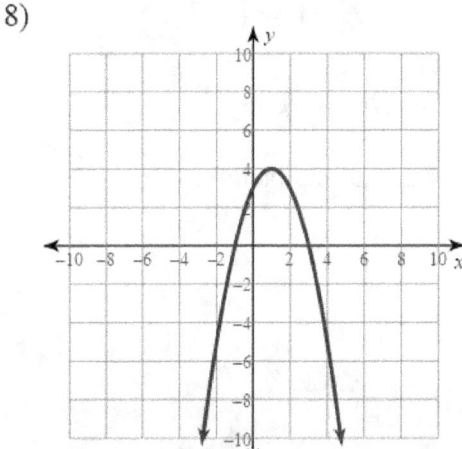

Identify the zeros:

What are the factors:

9)

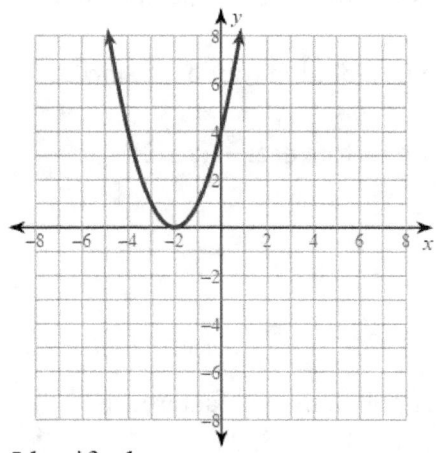

Identify the zeros:

What are the factors:

10)

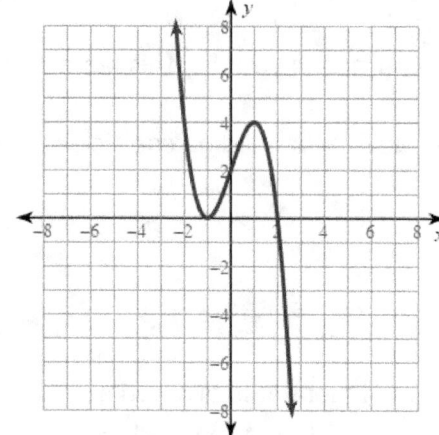

Identify the zeros:

What are the factors:

11)

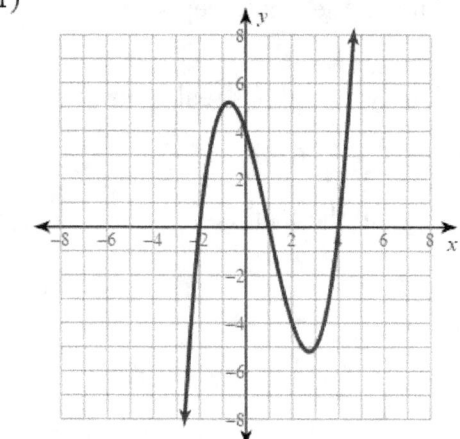

Identify the zeros:

What are the factors:

12)

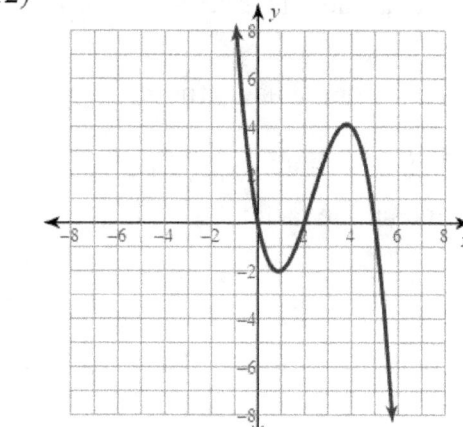

Identify the zeros:

What are the factors:

13)

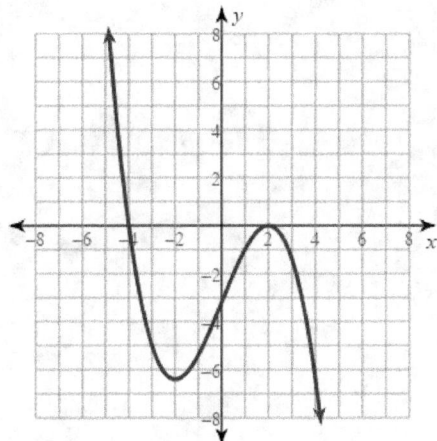

Identify the zeros:

What are the factors:

14)

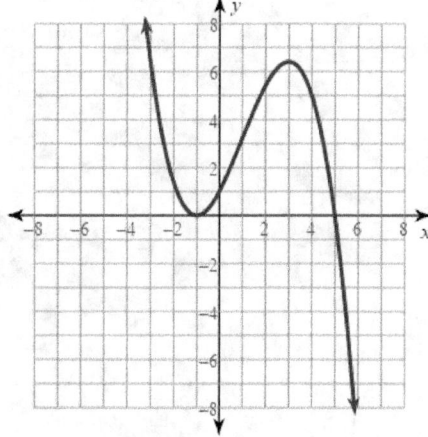

Identify the zeros:

What are the factors:

15)

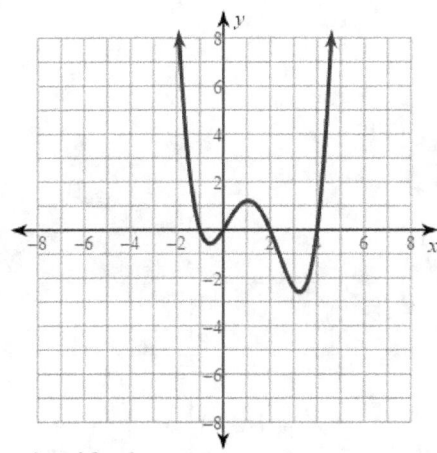

Identify the zeros:

What are the factors:

16)

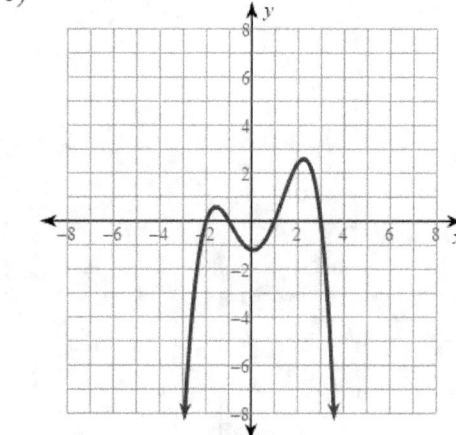

Identify the zeros:

What are the factors:

17)

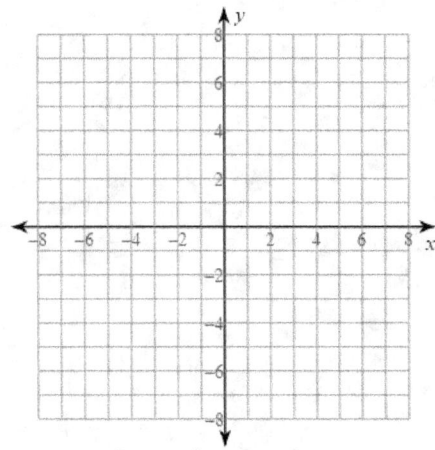

Graph $x^3 - 5x^2 + 2x + 8$

Identify the zeros:

What are the factors:

18)

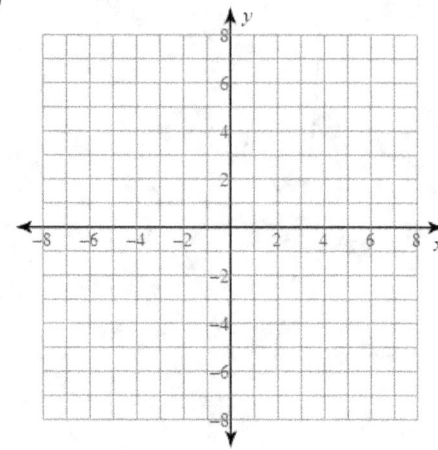

Graph $x^3 - 7x + 6$

Identify the zeros:

What are the factors:

19)

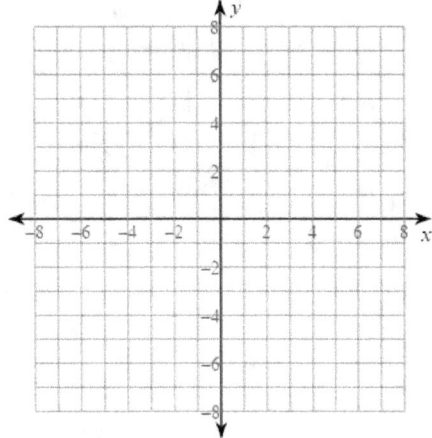

Graph $x^4 - 2x^3 - 9x^2 + 2x + 8$

Identify the zeros:

What are the factors:

20)

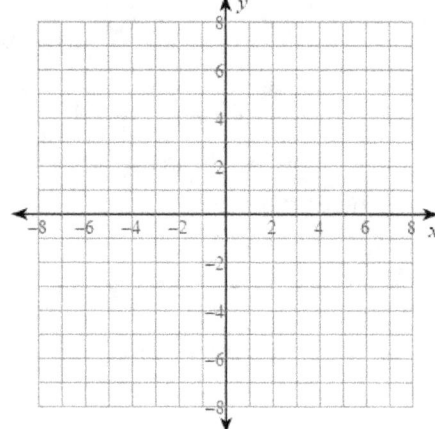

Graph $x^4 + 7x^3 + 8x^2 - 28x - 48$

Identify the zeros:

What are the factors:

21)

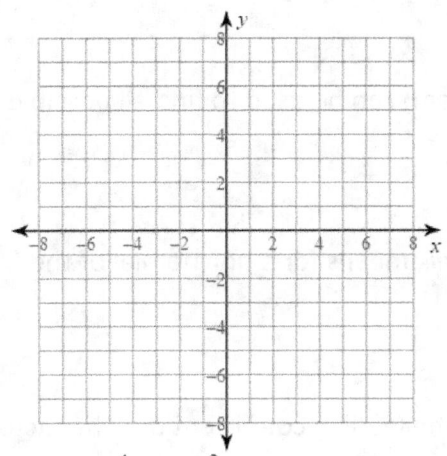

Graph $x^4 - 12x^2 + 16x$

Identify the zeros:

What are the factors:

22)

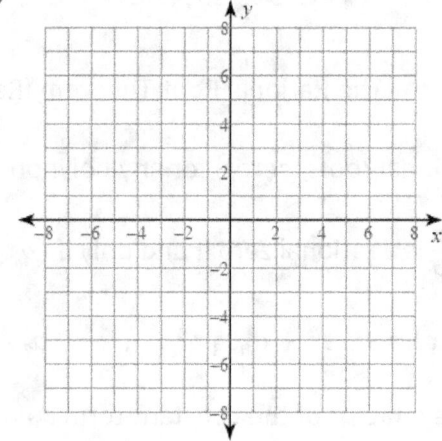

Graph $x^4 - 3x^3 - 9x^2 - 5x$

Identify the zeros:

What are the factors:

23)

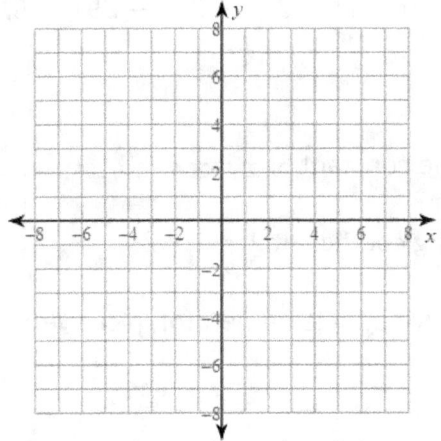

Graph $-x^5 + 5x^4 - 5x^3 - 5x^2 + 6x$

Identify the zeros:

What are the factors:

24)

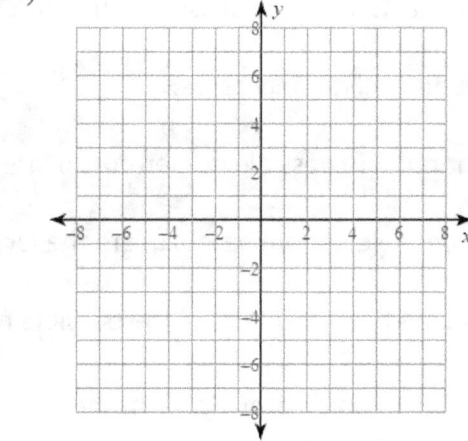

Graph $x^5 - 2x^4 - 7x^3 + 8x^2 + 12x$

Identify the zeros:

What are the factors:

Rational Root Theorem

The Rational Root Theorem (Rational Zero Theorem) can be used to find all possible rational roots (zeros) for any polynomial.

If $\frac{p}{q}$ is a rational zero if and only if $(qx - p)$ is a factor. This means for a polynomial of the form:

$$P(x) = a_n x^n + a_{n-1} x^{n-1} + \cdots + a_1 x + a_0$$

p is a factor of the constant term a_0 and q is a factor of the leading coefficient a_n. The factors can be either positive or negative.

After we know the possible rational zeros, then we can use synthetic division to find the actual rational zeros.

Example 1: State the possible rational zeros for the function $f(x) = 2x^3 + 3x^2 - 29x + 6$. Then find all rational zeros.

Solution: The possible numerators p, are the factors of the constant 6: $\pm 1, \pm 2, \pm 3, \pm 6$

The possible denominators q, are the factors of the leading coefficient 2: $\pm 1, \pm 2$

Combine these to make all the possible fractions. For every possible numerator, pick every possible denominator. Some of these will be duplicated, such as $\frac{1}{1} = \frac{-1}{-1} = \frac{2}{2} = \frac{-2}{-2}$.

The possible rational roots are $\pm 1, \pm 2, \pm 3, \pm 6, \pm \frac{1}{2}, \pm \frac{3}{2}$

Now use synthetic division to test which ones are roots.

Dividing by (x − 1)

$$\begin{array}{r|rrrr} 1 & 2 & 3 & -29 & 6 \\ & & 2 & 5 & -24 \\ \hline & 2 & 5 & -24 & \boxed{-18} \end{array}$$

Dividing by (x + 1)

$$\begin{array}{r|rrrr} -1 & 2 & 3 & -29 & 6 \\ & & -2 & -1 & 30 \\ \hline & 2 & 1 & -30 & \boxed{36} \end{array}$$

Dividing by (x − 2)

$$\begin{array}{r|rrrr} 2 & 2 & 3 & -29 & 6 \\ & & 4 & 14 & -30 \\ \hline & 2 & 7 & -15 & \boxed{-24} \end{array}$$

Dividing by (x + 2)

$$\begin{array}{r|rrrr} -2 & 2 & 3 & -29 & 6 \\ & & -4 & 2 & 54 \\ \hline & 2 & -1 & -27 & \boxed{60} \end{array}$$

Dividing by (x − 3)

$$\begin{array}{r|rrrr} 3 & 2 & 3 & -29 & 6 \\ & & 6 & 27 & -6 \\ \hline & 2 & 9 & -2 & \boxed{0} \end{array}$$

Remainder is zero, (x − 3) is a factor!

The other factor is $2x^2 + 9x - 2$

Dividing by (x + 3)

$$\begin{array}{r|rrrr} -3 & 2 & 3 & -29 & 6 \\ & & -6 & 9 & 60 \\ \hline & 2 & -3 & -20 & \boxed{66} \end{array}$$

There is a rational root at x = 3 and (x − 3) is a factor. Since we started with a cubic the remaining factor is quadratic, and we can use previous factoring techniques to break it down further to get the remaining factors and roots:

$$f(x) = 2x^3 + 3x^2 - 29x + 6 = (x - 3)(2x^2 + 9x - 2)$$

Find the discriminant of $2x^2 + 9x - 2$. $D = (9)^2 - 4(2)(-2) = 97$ So, the other roots will be irrational.

The only rational root is x = 3

Example 2: State the possible rational zeros for the function $f(x) = 3x^3 - 8x^2 - 22x + 45$.

Then find all rational zeros.

Solution: The possible numerators p, are the factors of the constant 45:

$\pm 1, \pm 3, \pm 5, \pm 9, \pm 15, \pm 45$

The possible denominators q, are the factors of the leading coefficient 3: $\pm 1, \pm 3$

Combine these to make all the possible fractions. For every possible numerator, pick every possible denominator. Some of these will be duplicated, such as $\frac{1}{1} = \frac{-1}{-1} = \frac{3}{3} = \frac{-3}{-3}$

The possible rational roots are $\pm 1, \pm 3, \pm 5, \pm 9, \pm 15, \pm 45, \pm \frac{1}{3}, \pm \frac{5}{3}$

Synthetic division can be quite a lengthy process and more difficult if the root is a fraction.

Graphing the function in Desmos is another option to find rational roots:

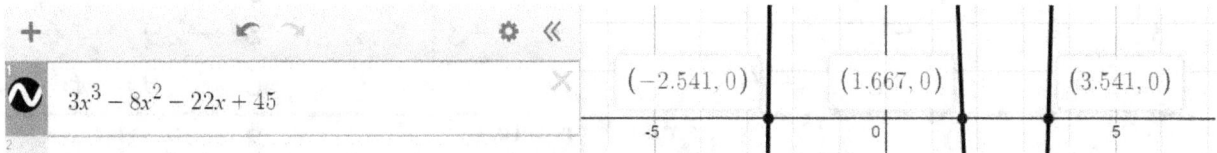

This cubic has three real roots, but the only possibilities for rational roots are:

$$\pm \frac{1}{3} = \pm 0.\overline{3} \text{ or } \pm \frac{5}{3} = \pm 1.\overline{6}$$

The decimals on Desmos are showing roots at $-2.541, 1.667, and\ 3.541$ so the only rational

root is $\frac{1}{3}$. The other two real roots must be irrational.

Practice Rational Root Theorem

State the possible rational zeros for each function.

1) $y = 2x^4 + 11x^2 - 6$

2) $y = 3x^5 + 9x^4 - 2x^3 - 6x^2 - 5x - 15$

3) $y = 2x^4 + 3x^2 - 27$

4) $y = 2x^4 + 21x^2 + 40$

5) $f(x) = 5x^4 + 4x^2 - 9$

6) $f(x) = 2x^5 + 10x^4 + 7x^3 + 35x^2 + 5x + 25$

7) $f(x) = 10x^5 + 2x^4 + 35x^3 + 7x^2 + 25x + 5$

8) $f(x) = 3x^4 - 10x^2 + 7$

9) $y = x^3 + 6x^2 - 21x - 18$

10) $y = 2x^3 + 5x^2 + 4x + 1$

11) $y = 5x^3 + x^2 - 5x - 1$

12) $y = x^3 - x^2 + 6x - 16$

State the possible rational zeros for each function. Then find all rational zeros.

13) $f(x) = 3x^3 + x^2 - 3x - 1$

14) $f(x) = 2x^3 - x^2 - 2x + 1$

15) $f(x) = 2x^3 + 5x^2 + 4x + 1$

16) $f(x) = 3x^3 - 10x^2 - 20x - 25$

17) $f(x) = x^3 - 12x^2 + 21x - 2$

18) $f(x) = x^3 - 15x^2 + 43x + 35$

19) $f(x) = 5x^3 + 21x^2 - 21x - 5$

20) $f(x) = x^3 + 11x^2 + 34x + 30$

Descartes' Rule of Signs

Descartes' Rule of Signs is a method of determining how many positive and negative real zeros a function can possibly have.

Let $P(x)$ be a polynomial in standard decreasing order, starting with highest exponent and going down, then:

1. The number of positive real zeros of $P(x)$ is equal to the number of variations in sign of $P(x)$, or that number decreased by an even amount.

2. The number of negative real zeros of $P(x)$ is equal to the number of variations in sign of $P(-x)$, or that number decreased by an even amount.

Example 1: State the possible number of positive and negative zeros for the function:

$$4x^5 - 10x^4 - 6x^3 + 15x^2 - 28x + 70$$

Solution: Count how many times the sign switches (Remember the first term here is positive):

$$+4x^5 - 10x^4 - 6x^3 + 15x^2 - 28x + 70$$
$$\quad 1 \qquad\qquad 2 \quad 3 \quad 4$$

There are 4 sign switches, so the number of positive real zeros can be 4 or decreased by an even amount 2, or 0.

Now evaluate the polynomial at –x:

$$4(-x)^5 - 10(-x)^4 - 6(-x)^3 + 15(-x)^2 - 28(-x) + 70$$

This will switch the sign of every term that has an odd exponent, and the sign will stay the same for every term with an even exponent:

$$-4x^5 - 10x^4 + 6x^3 + 15x^2 + 28x + 70$$

Descartes' Rule of Signs

Count how many times the sign switches.

$$-4x^5 - 10x^4 + 6x^3 + 15x^2 + 28x + 70$$

There is only 1 sign switch, so the number of negative real zeros can only be 1.

Possible # positive real zeros: 4, 2, or 0

Possible # negative real zeros: 1

Practice Descartes' Rule of Signs

State the possible number of positive and negative zeros for each function.

1) $f(x) \neq x^2 - 10x + 3$ p5. Zeros: 2, 0

$f(-x) = x^2 + 16x + 3$ neg. Zeros: 0

2) $f(x) = 3x^2 + x - 10$

3) $f(x) = 3x^3 + 5x^2 - 9x - 15$

4) $f(x) = 6x^3 + 31x^2 - 47x + 12$

5) $f(x) = 5x^4 + 14x^2 - 3$

6) $f(x) = 5x^4 - 12x^3 - 16x^2 + 8x$

7) $f(x) = 9x^5 - 15x^4 - 33x^3 + 55x^2 - 126x + 210$

8) $f(x) = 15x^5 + 25x^4 + 87x^3 + 145x^2 - 126x - 210$

9) $f(x) = 6x^6 + 10x^5 - 21x^4 - 35x^3 - 45x^2 - 75x$

10) $f(x) = 2x^6 - 6x^5 + 7x^4 - 21x^3 - 15x^2 + 45x$

11) $f(x) = 25x^7 + 100x^5 - 4x^3 - 16x$

12) $f(x) = 15x^7 + 5x^6 - 33x^5 - 11x^4 + 6x^3 + 2x^2$

13) $f(x) = 9x^8 - 148x^4 + 64$

14) $f(x) = 16x^8 + 80x^6 - x^4 - 5x^2$

15) $f(x) = 27x^9 - x^6 - 27x^3 + 1$

16) $f(x) = 27x^9 + 8x^6 - 27x^3 - 8$

17) $f(x) = 8x^{10} - x^7 - 8x^4 + x$

18) $f(x) = 27x^{10} + 8x^7 - 27x^4 - 8x$

19) $f(x) = 2x^5 - 4x^4 + 7x^3 - 14x^2 - 15x + 30$

20) $f(x) = 15x^4 - x^3 + 18x^2 - 8x$

Fundamental Theorem of Algebra

The Fundamental Theorem of Algebra states that a polynomial of degree n will have

exactly n complex roots (zeros) counting multiplicity (this includes real roots, imaginary roots,

and multiplicity of roots).

Conjugate Pair Theorem: If $a + bi$ is a complex zero of a polynomial, then $a - bi$ is also a complex zero. Imaginary roots **ALWAYS** come in conjugate pairs. Irrational roots also **ALWAYS** come in conjugate pairs $a + b\sqrt{c}$ and $a - b\sqrt{c}$.

Complex roots include real roots and imaginary roots. The number of possible real roots plus the number of possible imaginary roots equals the number of complex roots.

$$\begin{matrix} \textbf{Possible \# of} \\ \textbf{Real Roots} \end{matrix} + \begin{matrix} \textbf{Possible \# of} \\ \textbf{Imaginary Roots} \end{matrix} = \begin{matrix} \textbf{Total \# of} \\ \textbf{Complex Roots} \end{matrix} = \begin{matrix} \textbf{Degree of} \\ \textbf{Polynomial} \end{matrix}$$

Example 1: Given the function $f(x) = x^7 - 2x^6 + 5x^5 - 12x^4 - 13x^3 + 14x^2 + 7x$, state the number of complex zeros and the possible number of real and imaginary zeros. Also find the other roots given that two roots are: $1 + \sqrt{2}$ and $-i\sqrt{7}$:

Solution: The number of complex roots is the total number of roots and is equal to the degree of the polynomial. In this case the number of complex roots is 7. Imaginary roots come in pairs, so think about the most there could be if the number has to be even. The most is 6 and they have to go down by 2. So, the number of imaginary roots = 6, 4, 2, 0. The number of real roots added to the number of imaginary roots is equal to the total number of complex roots. Take 7 minus each number of possible imaginary roots. The number of real roots = 1, 3, 5, 7. The conjugate roots of the ones given are: $1 - \sqrt{2}$ and $i\sqrt{7}$.

Practice Fundamental Theorem of Algebra

1) $f(x) = x^3 - x + 6$

   # of complex zeros: **3**

   Possible # of real zeros: **1, 3**

   Possible # of imaginary zeros: **2, 0**

   If $1 + i\sqrt{2}$ is a root, then another root is:

       $1 - i\sqrt{2}$

2) $f(x) = x^3 + 64$

   # of complex zeros:

   Possible # of real zeros:

   Possible # of imaginary zeros:

   If $2 + 2i\sqrt{3}$ is a root, then another root is:

3) $f(x) = 5x^4 + 7x^2 + 2$

   # of complex zeros:

   Possible # of real zeros:

   Possible # of imaginary zeros:

   If $\dfrac{i\sqrt{10}}{5}$ is a root, then another root is:

   If i is a root, then another root is:

4) $f(x) = 5x^5 + 15x^4 + 8x^3 + 24x^2 + 3x + 9$

   # of complex zeros:

   Possible # of real zeros:

   Possible # of imaginary zeros:

   If $-\dfrac{i\sqrt{15}}{5}$ is a root, then another root is:

   If i is a root, then another root is:

5) $f(x) = 4x^5 - 5x^3 - 50x^2$

   # of complex zeros:

   Possible # of real zeros:

   Possible # of imaginary zeros:

   If $\dfrac{-5 - i\sqrt{55}}{4}$ is a root, then another root is:

6) $f(x) = 5x^4 + 43x^2 + 24$

   # of complex zeros:

   Possible # of real zeros:

   Possible # of imaginary zeros:

   If $-2i\sqrt{2}$ is a root, then another root is:

   If $\dfrac{i\sqrt{15}}{5}$ is a root, then another root is:

7) $f(x) = x^5 - 5x^4 - 39x^3 + 133x^2 - 40x + 138$

   # of complex zeros:

   Possible # of real zeros:

   Possible # of imaginary zeros:

   If $1 + \sqrt{47}$ is a root, then another root is:

   If i is a root, then another root is:

8) $f(x) = x^6 + 2x^5 - 20x^3 - 76x^2 + 48x + 240$

   # of complex zeros:

   Possible # of real zeros:

   Possible # of imaginary zeros:

   If $-1 + 3i$ is a root, then another root is:

   If $\sqrt{6}$ is a root, then another root is:

9) $f(x) = x^8 + 6x^7 - x^6 - x^2 - 6x + 1$

   # of complex zeros:

   Possible # of real zeros:

   Possible # of imaginary zeros:

   If $-3 + \sqrt{10}$ is a root, then another root is:

   If $\dfrac{-1 + i\sqrt{3}}{2}$ is a root, then another root is:

10) $f(x) = x^6 - 4x^5 - 30x^4 + 4x^3 + 27x^2 + 8x + 58$

   # of complex zeros:

   Possible # of real zeros:

   Possible # of imaginary zeros:

   If $2 + \sqrt{33}$ is a root, then another root is:

   If $-\sqrt{2}$ is a root, then another root is:

   If $-i$ is a root, then another root is:

11) $f(x) = x^{11} + 2x^{10} - 36x^9 + 8x^8 + 16x^7 - 288x^6 - x^5 - 2x^4 + 36x^3 - 8x^2 - 16x + 288$

   # of complex zeros:

   Possible # of real zeros:

   Possible # of imaginary zeros:

   If $-1 + \sqrt{37}$ is a root, then another root is:

   If $1 - i\sqrt{3}$ is a root, then another root is:

   If $\dfrac{-1 + i\sqrt{3}}{2}$ is a root, then another root is:

12) $f(x) = x^{10} - 6x^9 + 10x^8 - 26x^6 + 156x^5 - 260x^4 + 25x^2 - 150x + 250$

# of complex zeros:

Possible # of real zeros:

Possible # of imaginary zeros:

If $3 + i$ is a root, then another root is:

If $-i$ is a root, then another root is:

If $\sqrt{5}$ is a root, then another root is:

If $-i\sqrt{5}$ is a root, then another root is:

Factoring Higher Degrees

Now that we have developed all the factoring methods and discussed properties of polynomials, you are able to solve polynomial equations and graph polynomials.

Example 1: Find all roots of $f(x) = (2x - 1)(x + 5)^2(x - 7)(x^2 + 2)(x^2 - 3)(4x^2 - 2x + 1)$

Solution: This polynomial is in factored form with 4 linear factors (the x+5 factor counts twice) and 3 quadratic factors. Adding up all the degrees of the factors gives 10, so we should end up with 10 solutions. Set each of these to zero to find all the roots/zeros.

$(2x - 1) = 0 \Rightarrow x = \frac{1}{2}$ rational root

$(x + 5)^2 = 0 \Rightarrow x = -5$, with multiplicity 2 (double root)

$(x - 7) = 0 \Rightarrow x = 7$ rational root

$(x^2 + 2) = 0 \Rightarrow x^2 = -2 \Rightarrow x = \pm\sqrt{2}i$, 2 imaginary roots

Factoring Higher Degrees

$(x^2 - 3) = 0 \Rightarrow x^2 = 3 \Rightarrow x = \pm\sqrt{3}$, 2 irrational roots

$4x^2 - 2x + 1 = 0$ use quadratic formula to give $x = \frac{1 \pm \sqrt{3}i}{4}$, 2 imaginary roots

So, we have found all 10 roots: $\frac{1}{2}, -5$ $(mult.\ 2), 7, \pm\sqrt{2}i, \pm\sqrt{3}, \frac{1 \pm \sqrt{3}i}{4}$

Example 2: Factor and find all roots of $f(x) = 2x^6 - x^4 - 50x^2 + 25$

Solution: The polynomial is degree six, so it will have six solutions. Since this has 4 terms, try factoring by grouping:

$2x^6 - x^4 - 50x^2 + 25 = x^4(2x^2 - 1) - 25(2x^2 - 1) = (x^4 - 25)(2x^2 - 1)$

$$= (x^2 + 5)(x^2 - 5)(2x^2 - 1)$$

Set each factor to zero and solve:

$(x^2 + 5) = 0 \Rightarrow x^2 = -5 \Rightarrow x = \pm\sqrt{5}i$

$(x^2 - 5) = 0 \Rightarrow x^2 = 5 \Rightarrow x = \pm\sqrt{5}$

$(2x^2 - 1) = 0 \Rightarrow x^2 = \frac{1}{2} \Rightarrow x = \pm\sqrt{\frac{1}{2}} = \pm\frac{1}{\sqrt{2}} = \pm\frac{\sqrt{2}}{2}$

Example 3: Factor and find all roots of $f(x) = 64x^6 - 1$

Solution: Degree 6 polynomial equation will have 6 roots.

Factor using difference of squares:

$64x^6 - 1 = (8x^3 + 1)(8x^3 - 1)$

Each of these can be factored as a sum/difference of cubes.

$(8x^3 + 1)(8x^3 - 1) = (2x + 1)(4x^2 - 2x + 1)(2x - 1)(4x^2 + 2x + 1)$

Set the linear terms to zero:

$$(2x + 1) = 0 \Longrightarrow x = -\frac{1}{2} \; or \; (2x - 1) = 0 \Longrightarrow x = \frac{1}{2}$$

These quadratics cannot be factored, so solve each quadratic equation using the quadratic

formula:

$$4x^2 - 2x + 1 = 0 \Longrightarrow x = \frac{2 \pm \sqrt{(-2)^2 - 4(4)(1)}}{2(4)} = \frac{2 \pm \sqrt{-12}}{8} = \frac{2 \pm 2\sqrt{3}i}{8} = \frac{1 \pm \sqrt{3}i}{4}$$

$$4x^2 + 2x + 1 = 0 \Longrightarrow x = \frac{-2 \pm \sqrt{(2)^2 - 4(4)(1)}}{2(4)} = \frac{-2 \pm \sqrt{-12}}{8} = \frac{-2 \pm 2\sqrt{3}i}{8} = \frac{-1 \pm \sqrt{3}i}{4}$$

The roots: $x = \pm\frac{1}{2}, \frac{1\pm\sqrt{3}i}{4}, \frac{-1\pm\sqrt{3}i}{4}$

Example 4: Factor and find all roots of $2x^8 - 43x^6 - 64x^5 + 55x^4 + 96x^3 + 100x^2 + 160x = $

0. Given that $(1 + \sqrt{17})$ is a root.

Solution: Degree 8 polynomial equation will have 8 roots. Since $1 + \sqrt{17}$ is a root, $1 - \sqrt{17}$ is a

root. This means that $\left(x - \left(1 + \sqrt{17}\right)\right) and \left(x - \left(1 - \sqrt{17}\right)\right)$. Multiply these factors together

to get $\left(x - 1 - \sqrt{17}\right)\left(x - 1 + \sqrt{17}\right) = x^2 - x + \sqrt{17}x - x + 1 - \sqrt{17} - \sqrt{17}x + \sqrt{17} - 17$

$x^2 - x + \sqrt{17}x - x + 1 - \sqrt{17} - \sqrt{17}x + \sqrt{17} - 17 = x^2 - 2x - 16$

A GCF of x can be factored out first. Then use long division to divide $2x^7 - 43x^5 - 64x^4 + $

$55x^3 + 96x^2 + 100x + 160$ polynomial by $x^2 - 2x - 16$ to find the other factors:

$$x^2 - 2x - 16 \overline{\smash{\big)}\ \begin{array}{c} 2x^5 + 4x^4 - 3x^3 - 6x^2 - 5x - 10 \\ \hline 2x^7 + 0x^6 - 43x^5 - 64x^4 + 55x^3 + 96x^2 + 100x + 160 \end{array}}$$

$$\underline{-2x^7 + 4x^6 + 32x^5}$$
$$4x^6 - 11x^5 - 64x^4$$
$$\underline{-4x^6 + 8x^5 + 64x^4}$$
$$-3x^5 + 0x^4 + 55x^3$$
$$\underline{3x^5 - 6x^4 - 48x^3}$$
$$-6x^4 + 7x^3 + 96x^2$$
$$\underline{+6x^4 - 12x^3 - 96x^2}$$
$$-5x^3 + 0x^2 + 100x$$
$$\underline{+5x^3 - 10x^2 - 80x}$$
$$-10x^2 + 20x + 160$$
$$\underline{10x^2 - 20x - 160}$$
$$0$$

So, the other factor is $2x^5 + 4x^4 - 3x^3 - 6x^2 - 5x - 10$

So far, we have it factorized as: $x(x^2 - 2x - 16)(2x^5 + 4x^4 - 3x^3 - 6x^2 - 5x - 10)$

Use synthetic division to find another rational root of checking possible rational roots at

$$\pm 1, \pm 2, \pm 5, \pm 10, \pm \frac{1}{2}, \pm \frac{5}{2}$$

Dividing by (x + 2)

$$
\begin{array}{r|rrrrrr}
-2 & 2 & 4 & -3 & -6 & -5 & -10 \\
   &   & -4 & 0 & 6 & 0 & 10 \\
\hline
   & 2 & 0 & -3 & 0 & -5 & 0
\end{array}
$$

Remainder is zero,
x = -2 is a root
(x + 2) is a factor!

The other factor is $2x^4 - 3x^2 - 5$

we have it factorized as: $x(x^2 - 2x - 16)(x + 2)(2x^4 - 3x^2 - 5)$

Last thing to factor is $2x^4 - 3x^2 - 5$ which is of quadratic form and can be factored:

$$2x^4 - 3x^2 - 5 = (2x^2 - 5)(x^2 + 1)$$

The complete factorization is $x(x^2 - 2x - 16)(x + 2)(2x^2 - 5)(x^2 + 1)$

It has roots at $x = 0, 1 \pm \sqrt{17}, -2, \pm \frac{\sqrt{10}}{2}, \pm i$

Practice Factoring Higher Degrees

Find all roots.

1) $(x^2 - 5)(x^2 + 5)(x^2 - 2)(x^2 + 2) = 0$

$x^2 - 5 = 0 \quad x^2 + 5 = 0 \quad x^2 - 2 = 0 \quad x^2 + 2 = 0$

$x = \pm\sqrt{5}, \pm\sqrt{5}i, \pm\sqrt{2}, \pm\sqrt{2}i$

2) $(x - 1)(x + 1)(x^2 + 6) = 0$

3) $(x^2 - 5)^2 = 0$

4) $x^2(x - 4)(x + 2) = 0$

5) $(x^2 + 7)(x^2 - 2) = 0$

6) $(x^2 + 8)(x^2 + 1) = 0$

7) $x^2(x^2 + 2x - 25) = 0$

8) $(x - 1)(x + 1)(x^2 + 1)(x^2 - 5)(x^2 + 5) = 0$

9) $(x - 1)^2 \cdot (x + 1)^2 \cdot (x^2 + 1)^2 = 0$

10) $(x^2 + 2)(x^2 - 7) = 0$

11) $(x - 3)(x + 3)(x^2 - 7) = 0$

12) $(x^2 + 4)(x^2 + 2) = 0$

Factor each and find all roots.

13) $x^2 + x - 6 = 0$

14) $x^2 + 4x + 20 = 0$

15) $x^3 - 4x^2 - 11x = 0$

16) $x^3 - 2x^2 + 17x = 0$

17) $x^4 + 11x^2 + 30 = 0$

18) $x^4 - 2x^2 - 35 = 0$

19) $x^4 - 7x^2 - 8 = 0$

20) $x^2 - 8x - 1 = 0$

21) $x^5 + 2x^4 - 4x^3 - 8x^2 = 0$

22) $x^3 + 2x^2 - 15x = 0$

23) $x^3 - 6x^2 - 28x = 0$

24) $x^4 + 14x^2 + 48 = 0$

25) $x^4 - 3x^2 - 54 = 0$

26) $x^6 - 1 = 0$

27) $x^5 - 2x^4 + 2x^3 - 4x^2 - 15x + 30 = 0$

28) $x^5 - 2x^4 + 6x^3 - 12x^2 - 16x + 32 = 0$

Find all roots. One root has been given.

29) $9x^8 - 18x^7 + 81x^6 + 18x^5 - 106x^4 + 32x^3 - 144x^2 - 32x + 160 = 0;\ 1 + 3i$

30) $8x^8 - 32x^7 - 24x^6 + 65x^5 - 260x^4 - 195x^3 + 8x^2 - 32x - 24 = 0;\ 2 + \sqrt{7}$

31) $16x^{10} - 32x^9 + 32x^8 - 25x^6 + 50x^5 - 50x^4 + 9x^2 - 18x + 18 = 0; \ 1 + i$

32) $8x^{11} + 64x^{10} + 136x^9 - x^8 - 8x^7 - 17x^6 - 8x^5 - 64x^4 - 136x^3 + x^2 + 8x + 17 = 0; \ -4 + i$

Graphing Polynomials

Using a graphing calculator or Desmos, we can graph polynomials to find approximate

real zeros, and approximate relative extrema. Relative extrema are the relative maxima (peaks)

and relative minima (valleys) of a graph.

Example 1: State the maximum number of turns the graph of the function

$f(x) = x^5 - 4x^3 + x$ could make. Then sketch the graph. State the number of real zeros.

Approximate each zero to the nearest tenth. Approximate the relative minima and relative

maxima to the nearest tenth.

Solution: This is degree 5, so the maximum number of turns (extrema) is one less than the

degree, so 4. Here's what the graph looks like on Desmos:

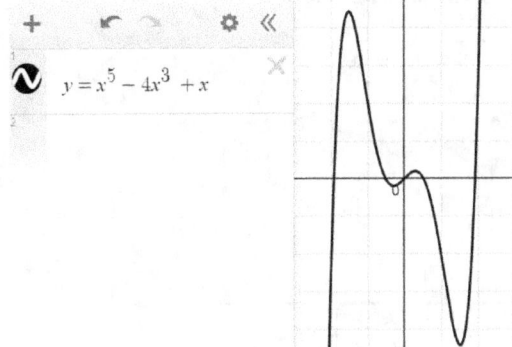

We can see it has 5 real zeros (corresponding to the x-intercepts). For this graph there is some

symmetry to the x-intercepts because the function is an odd function (all odd exponents). Click

on the graph and each x-intercept:

$(-1.932, 0)$ $(1.932, 0)$

$\bullet(0, 0)$

$(-0.518, 0)$ $(0.518, 0)$

$zeros: -1.9, -0.5, 0, 5, 1.9$

Click on the graph and the relative maxima and minima:

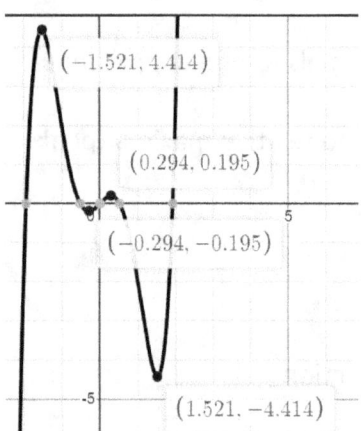

The relative max: $(-1.5, 4.4), (0.3, .2)$ and the relative min: $(-0.3, -0.2), (1.5, -4.4)$

Practice Graphing Polynomials

State the maximum number of turns the graph of each function could make. Then sketch the graph. State the number of real zeros. Approximate each zero to the nearest tenth. Approximate the relative minima and relative maxima to the nearest tenth.

1) $f(x) = -x^3 - 9x^2 - 24x - 16$

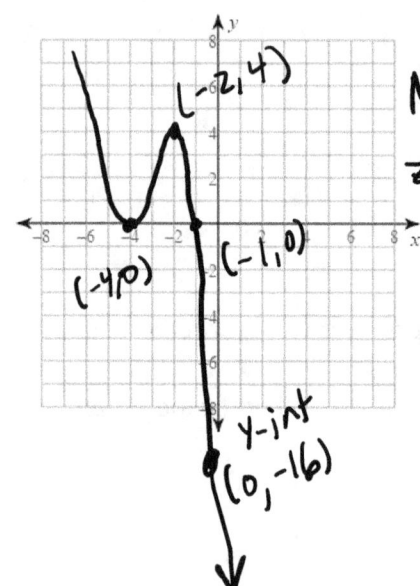

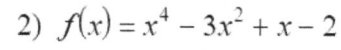

Max # Turns: 2

zeros: -4, -1

min: (-4, 0)

Max: (-2, 4)

2) $f(x) = x^4 - 3x^2 + x - 2$

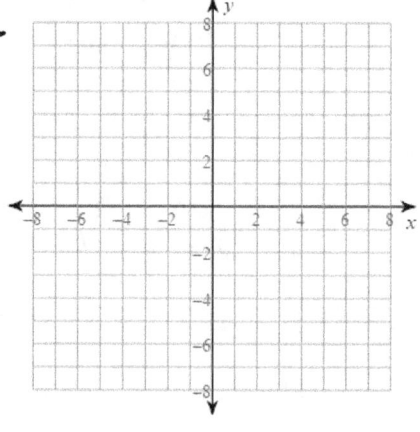

3) $f(x) = x^2 + 6x + 6$

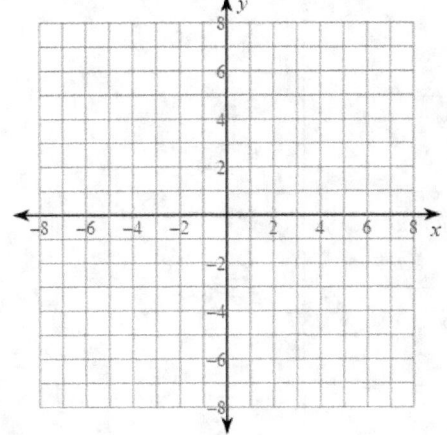

4) $f(x) = -x^3 + 3x^2 + 3$

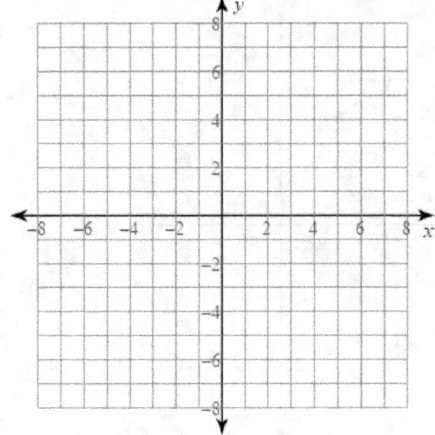

5) $f(x) = -x^5 + 4x^3 - x$

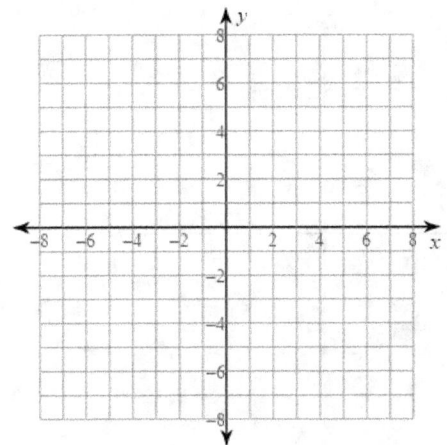

6) $f(x) = x^4 - x^2 + x - 4$

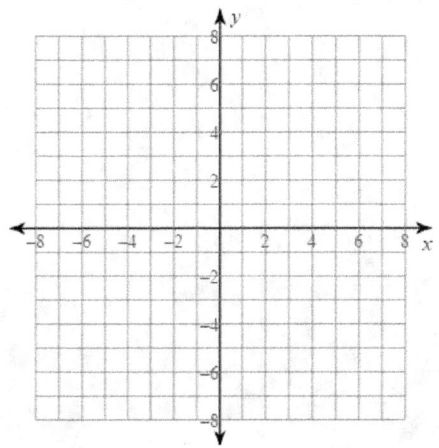

7) $f(x) = x^3 + 14x^2 + 60x + 79$

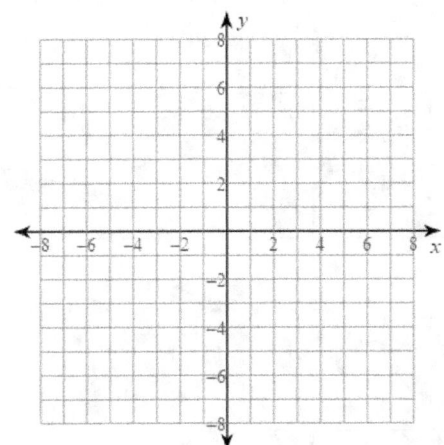

8) $f(x) = -x^2 - 1$

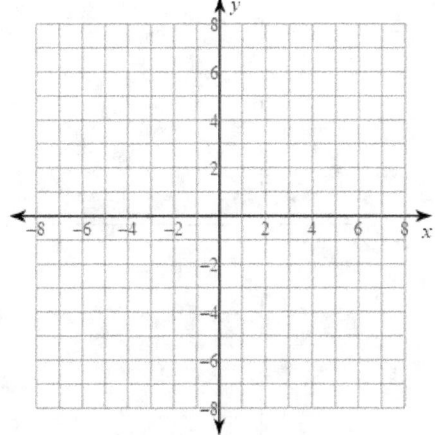

9) $f(x) = x^4 - 3x^2 - x + 3$

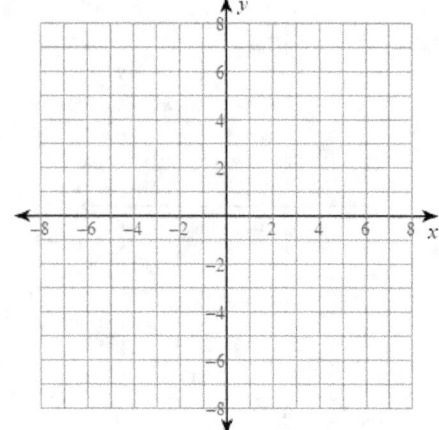

10) $f(x) = x^3 - 3x^2 - 1$

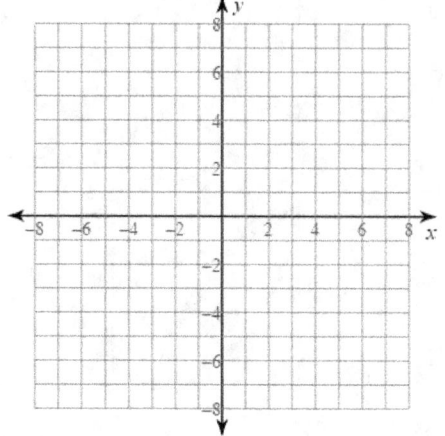

11) $f(x) = -x^4 + 3x^2 - 1$

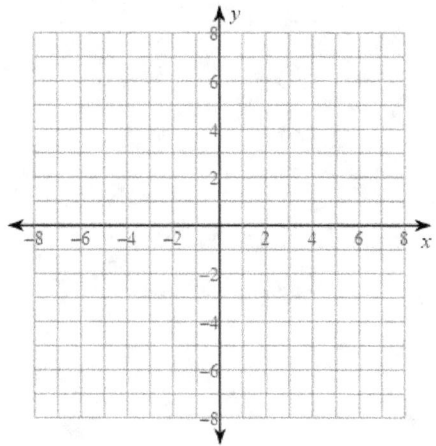

12) $f(x) = -x^4 + 4x^3 - 5x^2 + 3x + 1$

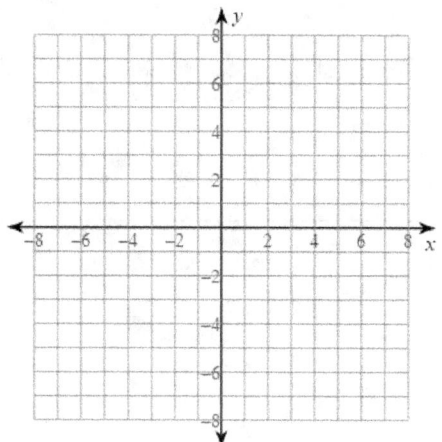

13) $f(x) = x^3 - 12x^2 + 45x - 53$

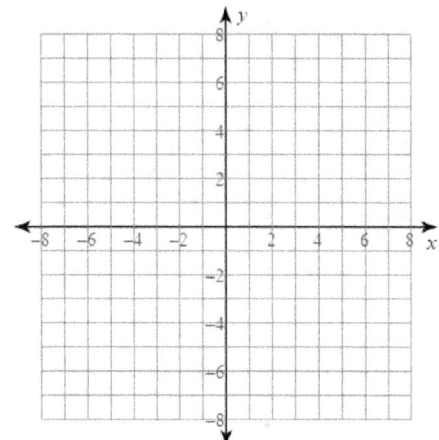

14) $f(x) = -x^3 + x^2 + 3$

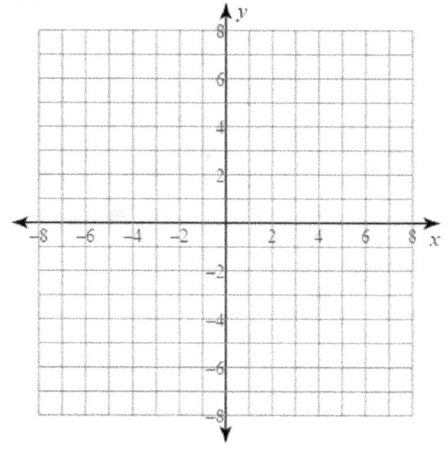

15) $f(x) = x^5 - 4x^3 + 3x - 4$

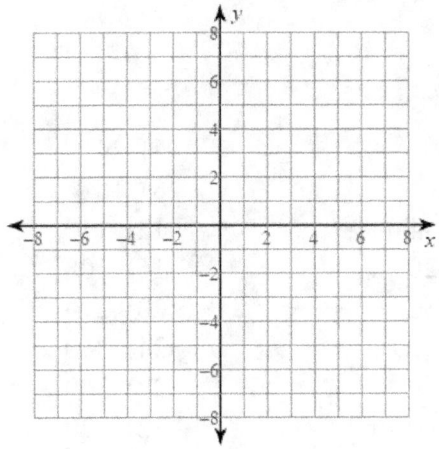

16) $f(x) = -x^3 + 3x^2 + 2$

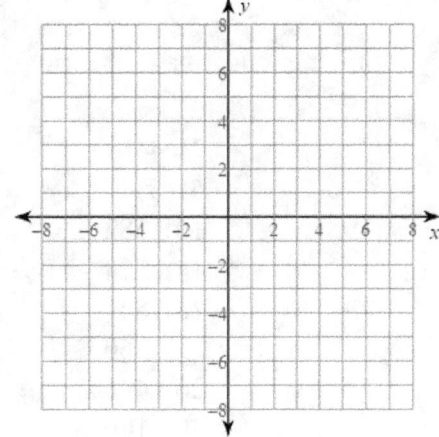

17) $f(x) = -x^5 + 3x^3 + 2$

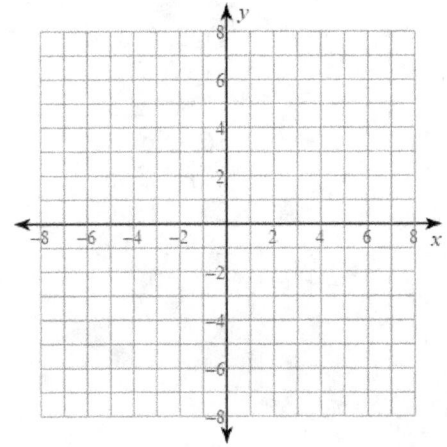

18) $f(x) = x^3 - 2x^2 - 3$

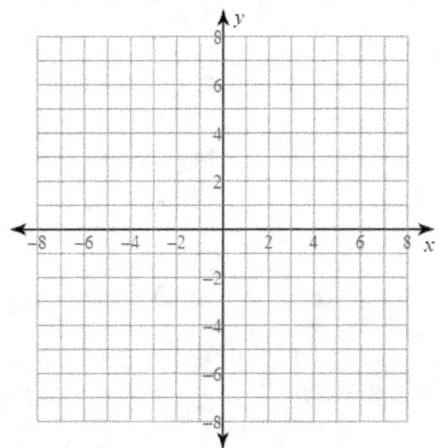

19) $f(x) = -x^4 + 2x^2 + x + 1$

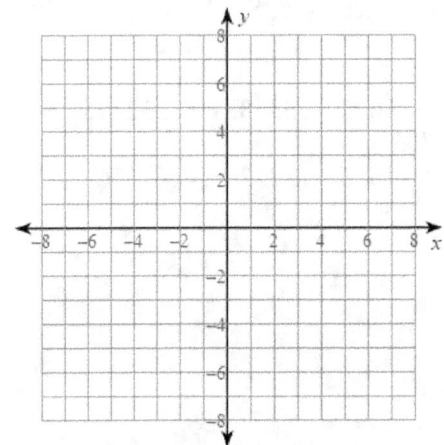

20) $f(x) = -x^4 + x^2 - x + 3$

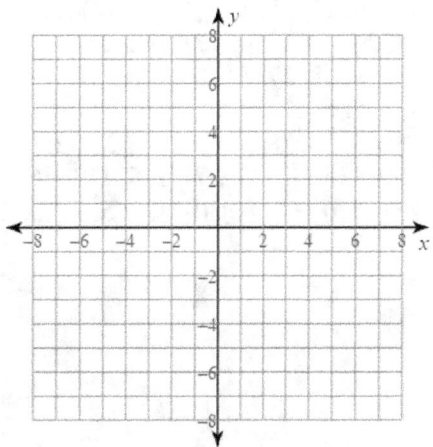

Solutions

Practice Multiplying Numbers Solutions

1) 18	2) 44	3) 130	4) 24
5) 30	6) 63	7) 42	8) 18
9) 96	10) 182	11) 144	12) 140
13) 200	14) 224	15) 120	16) 108
17) 36	18) 84	19) 280	20) 270
21) 840	22) 560	23) 1200	24) 840
25) 280	26) 448	27) 160	28) 360
29) 960	30) 1400	31) 480	32) 720

Practice Divisibility Solutions

1) 2, 3, 6	2) 3	3) 3, 5	4) 2, 3, 6, 9
5) 5	6) 2, 3, 6	7) None	8) None
9) None	10) 2	11) 2	12) None
13) 2	14) None	15) 3	16) None
17) 2, 5, 10	18) 2	19) 2	20) 2, 3, 5, 6, 10
21) 3, 9	22) 2	23) 2, 3, 6	24) 2, 3, 6
25) 2	26) None	27) None	28) 2
29) 3, 5	30) 5		

Practice Factors Solutions

1) 1, 2, 5, 10
2) 1, 2, 4, 8
3) 1, 2, 3, 6
4) 1, 3, 9
5) 1, 2, 3, 6, 9, 18
6) 1, 2, 3, 4, 6, 12
7) 1, 3, 5, 15
8) 1, 2, 4, 8, 16
9) 1, 2, 4, 5, 10, 20
10) 1, 2, 7, 14
11) 1, 2, 4, 7, 14, 28
12) 1, 2, 11, 22
13) 1, 2, 3, 5, 6, 10, 15, 30
14) 1, 5, 25
15) 1, 2, 3, 4, 6, 8, 12, 24
16) 1, 2, 13, 26
17) 1, 2, 23, 46
18) 1, 5, 11, 55
19) 1, 2, 31, 62
20) 1, 2, 3, 6, 9, 18, 27, 54
21) 1, 2, 5, 10, 25, 50
22) 1, 2, 3, 4, 5, 6, 10, 12, 15, 20, 30, 60
23) 1, 2, 4, 8, 16, 29, 32, 58, 116, 232, 464, 928
24) 1, 5, 13, 65, 169, 845
25) 1, 3, 5, 15, 53, 159, 265, 795
26) 1, 5, 167, 835
27) 1, 5, 113, 565
28) 1, 2, 4, 8, 16, 61, 122, 244, 488, 976
29) 1, 2, 263, 526
30) 1, 3, 5, 7, 15, 21, 35, 49, 105, 147, 245, 735

Practice Multiples Solutions

1) 2, 4, 6, 8, 10, 12, 14, 16, 18, 20
2) 3, 6, 9, 12, 15, 18, 21, 24, 27, 30
3) 4, 8, 12, 16, 20, 24, 28, 32, 36, 40
4) 5, 10, 15, 20, 25, 30, 35, 40, 45, 50
5) 6, 12, 18, 24, 30, 36, 42, 48, 54, 60
6) 7, 14, 21, 28, 35, 42, 49, 56, 63, 70
7) 8, 16, 24, 32, 40, 48, 56, 64, 72, 80
8) 9, 18, 27, 36, 45, 54, 63, 72, 81, 90
9) 10, 20, 30, 40, 50, 60, 70, 80, 90, 100
10) 11, 22, 33, 44, 55, 66, 77, 88, 99, 110

Practice Prime Factorization Solutions

1) $2^3 \cdot 3$
2) $2 \cdot 3^2$
3) $2 \cdot 7$
4) $3 \cdot 7$

5) 2^4
6) $2^2 \cdot 3$
7) $3 \cdot 5$
8) 5^2

9) $2^2 \cdot 5$
10) $2 \cdot 11$
11) $2 \cdot 3^2 \cdot 5$
12) $5 \cdot 11$

13) $3 \cdot 5^2$
14) $5 \cdot 17$
15) $2 \cdot 41$
16) $2^2 \cdot 3 \cdot 7$

17) $2 \cdot 29$
18) $2^4 \cdot 5$
19) $5 \cdot 13$
20) $2^2 \cdot 3 \cdot 5$

21) $3 \cdot 271$
22) $2^3 \cdot 5 \cdot 19$
23) $2^4 \cdot 41$
24) $3 \cdot 251$

25) $3 \cdot 239$
26) $2^2 \cdot 139$
27) $3 \cdot 193$
28) $2^2 \cdot 11 \cdot 19$

29) $7 \cdot 83$
30) $2 \cdot 3 \cdot 5^3$

Solutions

Practice Area of Rectangles Solutions

1) 15 km²	2) 42 mi²	3) 32 cm²	4) 63 cm²
5) 70 km²	6) 27 yd²	7) 98 m²	8) 144 in²
9) 140 km²	10) 54 yd²	11) 36 m²	12) 28 yd²
13) 7 m	14) 7 cm	15) 6 in	16) 9 m
17) 8 ft	18) 12 ft	19) 4 km	20) 8 in

21) 3
   1 x 12
   2 x 6
   3 x 4

22) 2
   1 x 14
   2 x 7

23) 1
   1 x 17

24) 3
   1 x 20
   2 x 10
   4 x 5

25) 4
   1 x 24
   2 x 12
   3 x 8
   4 x 6

26) 2
   1 x 25
   5 x 5

27) 4
   1 x 30
   2 x 15
   3 x 10
   5 x 6

28) 1
   1 x 31

29) 5
   1 x 48
   2 x 24
   3 x 16
   4 x 12
   6 x 8

30) 2
   1 x 49
   7 x 7

31) The area must be a prime number.

32) The area must be the square of a prime number.

Practice Least Common Multiple Solutions

1) 60	2) 18	3) 16	4) 24
5) 24	6) 30	7) 36	8) 42
9) 48	10) 30	11) 60	12) 84

Practice Greatest Common Factor Solutions

1) 6	2) 4	3) 2	4) 3
5) 1	6) 8	7) 3	8) 1
9) 3	10) 6	11) 5	12) 2

Practice LCM and GCF Using a Venn Diagram Solutions

1) LCM: 80
   GCF: 8

2) LCM: 60
   GCF: 10

3) LCM: 54
   GCF: 9

4) LCM: 80
   GCF: 4

5) LCM: 160
   GCF: 8

6) LCM: 30
   GCF: 5

7) LCM: 32
   GCF: 16

8) LCM: 90
   GCF: 6

9) LCM: 108
   GCF: 9

10) LCM: 120
    GCF: 10

11) LCM: 84
    GCF: 14

12) LCM: 240
    GCF: 12

13) LCM: 66
    GCF: 11

14) LCM: 210
    GCF: 5

15) LCM: 275
    GCF: 5

16) LCM: 280
    GCF: 7

17) LCM: 112
    GCF: 8

18) LCM: 168
    GCF: 14

19) LCM: 330
    GCF: 5

20) LCM: 110
    GCF: 11

21) LCM: 360
    GCF: 12

22) LCM: 72
    GCF: 4

23) LCM: 315
    GCF: 5

24) LCM: 660
    GCF: 11

25) LCM: 330
    GCF: 11

26) LCM: 600
    GCF: 5

27) LCM: 672
    GCF: 8

28) LCM: 72
    GCF: 4

29) LCM: 540
    GCF: 9

30) LCM: 1320
    GCF: 4

31) LCM: 216
    GCF: 9

32) LCM: 288
    GCF: 24

Practice Combining Like Terms Solutions

1) $-3 + 4m$

2) $-8a - 7$

3) $8 - 7n$

4) $3n + 11$

5) $10k + 2$

6) $8x - 8$

7) $7 - 9b$

8) $16x + 6$

9) 10

10) $4 + m$

11) $2 + 11x$

12) $13 - 3x$

13) $9 - 2x$

14) $14 + b$

15) $-m^2 + m + 14$

16) $6n - 8$

17) $2x^2 + 11x + 7$

18) $5x^2 + 2 + 2x$

19) $-2x^2 - 9x - 5$

20) $6x^2 - 12k + 20$

Solutions

Practice Properties of Exponents Solutions

1) 8^5

2) 8^7

3) -6

4) $\dfrac{1}{4^2}$

5) 2^{12}

6) 5^8

7) $\dfrac{1}{(-3)^4}$

8) $\dfrac{1}{4^9}$

9) $7x^3y^5$

10) $20x^4y^7$

11) $16a^6b^5$

12) $49x^9y^7$

13) $\dfrac{5x^3}{6y^3}$

14) $\dfrac{y^2}{8x}$

15) $\dfrac{6x}{5}$

16) $\dfrac{5u^2}{7v^2}$

17) $216x^9y^{12}$

18) $125n^{12}$

19) $25m^6n^4$

20) $64v^{12}$

21) $15x^4y^5$

22) $4u^2v^4$

23) $56x^2$

24) $\dfrac{18}{x^4y}$

25) $\dfrac{4}{m^6n^3}$

26) $\dfrac{2a^2b^7}{7}$

27) $\dfrac{4x^4y^2}{5}$

28) $\dfrac{4x^2}{7y^6}$

29) $\dfrac{1}{27b^9}$

30) $8u^{12}v^9$

31) $\dfrac{b^2}{25a^2}$

32) $\dfrac{8x^{12}}{y^{12}}$

33) $\dfrac{1}{81m^{28}n^{16}}$

34) $\dfrac{1}{2x^{15}y}$

35) 1

36) $\dfrac{x^2}{4y^3}$

37) 1

38) $\dfrac{1}{81x^4}$

39) $\dfrac{4b^2}{27a^{11}}$

40) $\dfrac{81b^{10}}{8a^5}$

Practice Factoring Monomials Solutions

1) $2 \cdot 2 \cdot 2r \cdot r \cdot s \cdot s \cdot s$ 2) $2 \cdot 11a \cdot a \cdot a \cdot a \cdot b$ 3) $2 \cdot 2 \cdot 2 \cdot 2x \cdot x \cdot y \cdot y \cdot y$

4) $2 \cdot 5r \cdot r \cdot s \cdot s \cdot s \cdot s \cdot s$ 5) $2 \cdot 3 \cdot 5x \cdot x \cdot y \cdot z \cdot z \cdot z$ 6) $3 \cdot 5m \cdot p \cdot p \cdot p \cdot q \cdot q$

7) $2 \cdot 2 \cdot 2 \cdot 3x \cdot x \cdot y \cdot y \cdot y \cdot y \cdot z \cdot z \cdot z$ 8) $3 \cdot 3x \cdot x \cdot y \cdot y \cdot y \cdot y \cdot z \cdot z \cdot z$

9) $2 \cdot 2 \cdot 3m \cdot m \cdot n \cdot n \cdot p \cdot p \cdot p$ 10) $2 \cdot 13x \cdot x \cdot x \cdot x \cdot y \cdot y \cdot z$

Practice Factoring GCF from Monomials Solutions

1) $2a^2$

2) $44x$

3) $9y$

4) $16x$

5) $24xy$

6) $6x^3$

7) $6x^2$

8) 2

9) $7yx^2$

10) $3ab$

11) $9y^2$

12) $6n^2$

13) $2x^2y$

14) $2x^2y^2$

15) $3uvw$

16) $44uw^2$

17) $24u^3vw^2$

18) $6m^2np^4$

19) $7x^3y^4$

20) $2vw^3$

Practice Distributive Property Solutions

1) $-40 + 48x$ 2) $56b - 49$ 3) $12r + 32$ 4) $-1 - 2n$

5) $20 + 4x$ 6) $-20 + 6x$ 7) $42n^2 + 48n$ 8) $35m^4 - 35m^3$

9) $12x^5 + 6x^4$ 10) $14p^5 + 14p^4$ 11) $25x - 5$ 12) $18x^3 + 24x^2$

13) $8k^2 - 8k + 10$ 14) $25p^2 - 10p - 35$ 15) $16m^2 - 16m - 8$ 16) $21k^2 + 35k + 42$

17) $2x^5 + 8x^4 - 16x^3$ 18) $15m^4 - 9m^3 - 12m^2$ 19) $36p^3 + 24p^2 - 12p$

20) $6n^5 - 15n^4 + 18n^3$ 21) $3n^7 - 12n^6 - 21n^5$ 22) $40a^4 - 35a^3 - 35a^2$

23) $45n + 30$ 24) $-62n - 6$ 25) $-31 + 74a$ 26) $-67n - 43$

27) $-20x - 8$ 28) $-52n + 2$ 29) $-1 + 23n$ 30) $78x - 39$

31) $-11 - 60n$ 32) $-31n + 25$ 33) $9a^2 - 3a - 20$ 34) $20p^2 + 13p - 21$

35) $28n^2 - 54n + 18$ 36) $35b^2 - 36b + 9$

Practice Distributive Property with Algebra Tiles Solutions

1) $2x + 4$ 2) $2x + 2$ 3) $2x - 6$ 4) $-2x + 6$

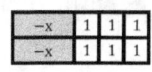

5) $6x - 4$ 6) $3x + 6$ 7) $2x - 4$ 8) $3x - 6$

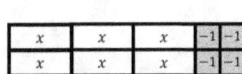

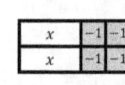

9) $9x + 6$ 10) $6x - 3$ 11) $6x + 6$ 12) $2 + 4x$

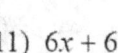

Practice Factor Out GCF Solutions

1) $7n(-3n + 5)$
5) $8n(7n - 3)$
9) $8n^3(6 + 10n + n^2)$

2) $8(-10 + 3v)$
6) $3r(5r + 7)$
10) $8b^2(5 - 2b - 6b^2)$

3) $7(x + 2)$
7) $9(4n^3 - 2n - 3)$
11) $6n^2 - 9n + 2$
There is no GCF other than 1

4) $2p^2(5p - 3)$
8) $9(7m^2 + 4m + 10)$

12) $10(3n^4 + 3n^3 - 2)$
15) $10y^3(3y^3 - 3xy + 7)$
18) $4x^2y(5xy^3 + 5y - 3)$
20) $7k(3k^3 + 5h^3k^4 + 3h^5j - 2j^3)$
22) $10a(9b^5 - 8b - a^4b - 5c^3)$
24) $5pr^2(2r^3 + 4pr - 7r^2 + 9q^2)$
26) $10(-7m^3np + 11)$
28) $8a^2b^3(ab^3 + 9a - 2b)$

13) $5x^7(6y^2 + 5x^2 + 8y)$
16) $4xy^6(5xy^2 + 6x + 5)$
19) $5x^2z(2x^2y^5z^2 - 8y^2z^4 + 9x^2z + y^2)$

21) $6x^3(-10y^4z + 5x^3z - 9z^2 + 2y)$
23) $6c^2(2b^4 - 2 + 10ac^2 + 7a^2c)$
25) $9n(2n^3 + 12n^2 - 9n + 7)$
27) $6x^2z(4x^7y^4z + 2x^3y^3 + 5y + 6)$
29) $6 - 10hk + 11j^2k + 14j^2$
GCF = 1
30) $6x(2y^3z^3 + 10z^2 - 5y)$

31) $49qr^8(-15p^2q^3 + 18qr - 20p^2 + 2q)$
32) $17y^4(-17x^3y^4z + 19y^3z^3 + 23x + 16)$

14) $x^3(7x^2y^3 + 2x^2y + 3)$
17) $2x^2y(4 + 4x - x^3)$

Practice Multiplying Binomials Leading Coefficient of 1 Solutions

1) $x^2 + 3x - 28$
5) $m^2 - m - 12$
9) $a^2 - 5a - 14$
12) $22m^2 - 126mn - 36n^2$
15) $m^2 - 16mn + 55n^2$
19) $p^2 + 10pq + 9q^2$

2) $x^2 - 9x + 18$
6) $b^2 - 9b + 20$
10) $n^2 - 2n - 99$

3) $n^2 + 10n + 16$
7) $x^2 + 16x + 55$
11) $x^2 + 9xy + 14y^2$
13) $p^2 - 17pq + 66q^2$
16) $b^2 - 22ab + 120a^2$
20) $x^2 + 15xy + 36y^2$

4) $n^2 - 3n - 10$
8) $n^2 + n - 72$

14) $a^2 - 14ab + 40b^2$
17) $r^2 - 6rs - 27s^2$

18) $m^2 - 5mn - 84n^2$

Practice Factoring Trinomials of the Form x^2+bx+c Solutions

1) $(x - 5)(x - 8)$
5) $(n + 5)(n - 8)$
9) $(v - 6)(v - 4)$
13) $(x - 10y)^2$
17) $(x - 6y)(x - 8y)$
21) $2(x + 10)(x - 4)$
25) $3(r - 3)(r - 7)$
29) $2(u + 6v)(u + 4v)$
32) $5(x - 11y)(x - 13y)$

2) $(n - 10)(n + 7)$
6) $(n + 1)(n + 10)$
10) $(x + 10)(x - 2)$
14) $(a + b)(a - 3b)$
18) $3(a + b)(a - b)$
22) $3m(m + 10)$
26) $2a(a - 4b)$
30) $5(u - v)(u - 5v)$

3) $(k - 8)(k - 2)$
7) $(r - 10)(r + 5)$
11) $6(x + 10y)(x - 5y)$
15) $(x + 8y)(x + 2y)$
19) $(u + 3v)(u + 6v)$
23) $5(m + 9)(m - 2)$
27) $2(x - y)(x + 10y)$
31) $2(m - 4n)(m - 12n)$

4) $(v + 5)(v + 4)$
8) $(x - 3)(x - 5)$
12) $(a + 8b)(a - 7b)$
16) $4(x - 9y)(x + 9y)$
20) $(x - 10y)(x - 7y)$
24) $5(x - 6)(x - 2)$
28) $5(a - 7b)(a + 8b)$

Practice Squares and Cubes Solutions

1) 1.7	2) 2.8	3) 2.2	4) 3.7
5) 2.6	6) 2.4	7) 3.2	8) 1.4
9) 4.7	10) 4.9	11) 4.4	12) 3.5
13) 5.1	14) 4.1	15) 5.6	16) 6.9
17) 5.8	18) 5.3	19) 7.7	20) 8.1
21) 7.5	22) 7.2	23) 9.5	24) 6.2
25) 8.7	26) 10.2	27) 6.6	28) 11.4
29) 13.8	30) 11.8	31) 13.2	32) 12.2
33) 16.1	34) 12.6	35) 14.8	36) 23.2
37) 20.5	38) 17.3	39) 19.5	40) 10.9

Practice Difference of Squares Pattern Solutions

1) $n^2 - 36$	2) $v^2 - 9$	3) $x^2 - 4$	4) $x^2 - 16$
5) $n^2 - 64$	6) $n^2 - 25$	7) $k^2 - 16$	8) $k^2 - 81$
9) $64m^2 - 25$	10) $25x^2 - 49$	11) $9v^2 - 25$	12) $4x^2 - 9$
13) $16a^2 - 1$	14) $64n^2 - 25$	15) $16p^2 - 36$	16) $36b^2 - 16$
17) $25x^2 - 64y^2$	18) $64a^2 - 64b^2$	19) $36x^2 - 4y^2$	20) $64y^2 - 25x^2$
21) $25x^2 - 25y^2$	22) $64x^2 - 4y^2$	23) $49y^2 - 36x^2$	24) $49x^2 - 64y^2$
25) $196x^2 - 36y^2$	26) $81x^2 - 361y^2$	27) $324y^2 - 289x^2$	28) $289u^2 - 25v^2$
29) $225x^2 - 324y^2$	30) $169x^2 - 64y^2$	31) $9x^2 - 169y^2$	32) $4y^2 - 225x^2$

Practice Factoring Difference of Squares Solutions

1) $(b + 10)(b - 10)$ 2) $(x + 7)(x - 7)$ 3) $(v + 5)(v - 5)$ 4) $(n + 6)(n - 6)$
5) Not factorable 6) $(b + 9)(b - 9)$ 7) $(x + 8)(x - 8)$ 8) $(n + 2)(n - 2)$
9) $(10a + 7b)(10a - 7b)$ 10) $(8x + 3y)(8x - 3y)$ 11) $(4x + 9y)(4x - 9y)$
12) $(2m + n)(2m - n)$ 13) $(8x + 9y)(8x - 9y)$ 14) $(7x + 10y)(7x - 10y)$
15) $(7m + 4n)(7m - 4n)$ 16) $(2x + 7y)(2x - 7y)$ 17) $(6m + n)(6m - n)$
18) $8(2x + 9y)(2x - 9y)$ 19) $(5u + 2v)(5u - 2v)$ 20) $10(2x + y)(2x - y)$
21) $(4a + 3b)(4a - 3b)$ 22) Not factorable 23) $6(5x + 7y)(5x - 7y)$
24) $(x + 9y)(x - 9y)$ 25) $(7x + 11y)(7x - 11y)$ 26) $(3x + 17y)(3x - 17y)$
27) $5(14x + 13y)(14x - 13y)$ 28) $2(14m + 11n)(14m - 11n)$ 29) $8(n + 3m)(n - 3m)$
30) $(5x + 12y)(5x - 12y)$ 31) $20(2x^2 + 15y^2)(2x^2 - 15y^2)$ 32) $10(169m^4 + 64n^4)$

Solutions

Practice Squaring Binomials Solutions

1) $x^2 + 16x + 64$
2) $v^2 + 2v + 1$
3) $v^2 - 14v + 49$
4) $a^2 - 16a + 64$
5) $b^2 - 6b + 9$
6) $b^2 + 14b + 49$
7) $p^2 + 10p + 25$
8) $b^2 - 12b + 36$
9) $4b^2 + 20b + 25$
10) $64x^2 - 80x + 25$
11) $121x^2 - 66x + 9$
12) $25b^2 - 70b + 49$
13) $81x^2 - 198x + 121$
14) $36m^2 + 84m + 49$
15) $4m^2 - 4m + 1$
16) $121p^2 - 154p + 49$
17) $100a^2 + 60ab + 9b^2$
18) $121x^2 - 264xy + 144y^2$
19) $25x^2 - 80xy + 64y^2$
20) $25u^2 + 120uv + 144v^2$
21) $25x^2 - 70xy + 49y^2$
22) $9y^2 + 6yx + x^2$
23) $16x^2 + 88xy + 121y^2$
24) $9x^2 + 60xy + 100y^2$
25) $4m^2 - 60mn + 225n^2$
26) $25v^2 + 80vu + 64u^2$
27) $25x^2 + 160xy + 256y^2$
28) $64y^2 + 144yx^2 + 81x^4$
29) $144x^2 - 168xy^2 + 49y^4$
30) $x^6 + 34x^3y^2 + 289y^4$
31) $144m^2 + 312mn^2 + 169n^4$
32) $49m^4 - 280m^2n^3 + 400n^6$

Practice Factoring Perfect Square Trinomials Solutions

1) $(3x - 10)^2$
2) $(9x + 1)^2$
3) $(7x - 2)^2$
4) $(5n + 2)^2$
5) $(5x + 4)^2$
6) $(8n + 7)^2$
7) $(7n - 1)^2$
8) $(6v - 5)^2$
9) $(10a - 3b)^2$
10) $(3x - 10y)^2$
11) $(7x + 5y)(7x - 5y)$
12) $(3a + 2b)^2$
13) $(8u - v)^2$
14) $4(7m + n)(7m - n)$
15) $7y(4x - 7y)^2$
16) $(7a + 3b)^2$
17) $(6a + 11b)^2$
18) $(9a - 19b)^2$
19) $14(18x - 19y)^2$
20) $3y(17x + 5y)^2$
21) $(8m + 17n)^2$
22) $6(17b - 14a)^2$
23) $(7x - 12y)^2$
24) $(18m - 13n)^2$
25) $(y^2 + x^2)^2$
26) $5(2y^2 + 5x^2)^2$
27) $2n(5x^2 + 4y^2)^2$
28) $3(3x^2 - y^2)^2$
29) $3(3m^3 + 2n^3)^2$
30) $(3u^3 - 2v^2)^2$
31) $(4x^2 - 5y^3)^2$
32) $3y^2(2x^3 - 3y^3)^2$

Practice Multiplying Polynomials More than Two Terms Solutions

1) $35r^3 + 50r^2 + 36r + 9$
2) $6x^3 - 24x^2 + 28x - 30$
3) $14k^3 + 38k^2 + 2k - 4$
4) $35n^3 + 31n^2 - 60n - 32$
5) $7u^3 + 45u^2v - 29uv^2 - 7v^3$
6) $56u^3 + 72u^2v + 32uv^2 + 16v^3$
7) $35x^3 - 5x^2y + 19xy^2 + 42y^3$
8) $4x^3 - 33x^2y + 37xy^2 - 14y^3$
9) $5b^4 - 26b^3 - 3b^2 + 44b - 20$
10) $20x^4 + 24x^3 - 69x^2 - 60x + 8$
11) $14k^4 + 23k^3 + 47k^2 + 51k - 35$
12) $4a^4 - 5a^3 - 38a^2 - 25a + 4$
13) $121a^4 - 110a^3b + 223a^2b^2 - 90ab^3 + 81b^4$
14) $4x^4 + 38x^3y + 58x^2y^2 - 126xy^3 + 42y^4$
15) $12x^4 - 14x^3y + 62x^2y^2 - 36xy^3 + 96y^4$
16) $12x^4 + 50x^3y + 56x^2y^2 + 20xy^3 - 6y^4$
17) $64x^3 + 27$
18) $64m^3 + 125$
19) $8u^3 - 125$
20) $27u^3 - 8$

Practice Factoring Sum/Difference of Cubes Solutions

1) $(3x + 5)(9x^2 - 15x + 25)$ 2) $(2x - 5)(4x^2 + 10x + 25)$ 3) $(3 - 5x)(9 + 15x + 25x^2)$
4) $(5u + 4)(25u^2 - 20u + 16)$ 5) $(3u - 2)(9u^2 + 6u + 4)$ 6) $(5u + 2)(25u^2 - 10u + 4)$
7) $(4x + 5)(16x^2 - 20x + 25)$ 8) $(x - 4)(x^2 + 4x + 16)$
9) $(9x - 4y)(81x^2 + 36xy + 16y^2)$ 10) $(5a + 7b)(25a^2 - 35ab + 49b^2)$
11) $(-10x + 7y)(100x^2 + 70xy + 49y^2)$ 12) $(2m - 9n)(4m^2 + 18mn + 81n^2)$
13) $(9u - 5v)(81u^2 + 45uv + 25v^2)$ 14) $(6m + 7n)(36m^2 - 42mn + 49n^2)$
15) $(9u + 4v)(81u^2 - 36uv + 16v^2)$ 16) $(6u - 5v)(36u^2 + 30uv + 25v^2)$
17) $(x - 10y)(x^2 + 10xy + 100y^2)$ 18) $y(x - 10y)(x^2 + 10xy + 100y^2)$
19) $y(9x + 7y)(81x^2 - 63xy + 49y^2)$ 20) $(10x - 7y)(100x^2 + 70xy + 49y^2)$
21) $(5m - 6n)(25m^2 + 30mn + 36n^2)$ 22) $x^3y(x + 9y)(x^2 - 9xy + 81y^2)$
23) $(7a + 10b)(49a^2 - 70ab + 100b^2)$ 24) $y^2(-9x - 4y)(81x^2 - 36xy + 16y^2)$
25) $3x(5x - 8y)(25x^2 + 40xy + 64y^2)$ 26) $x(3x + 7y)(9x^2 - 21xy + 49y^2)$
27) $(9x + 2y)(81x^2 - 18xy + 4y^2)$ 28) $6y(-5x - 6y)(25x^2 - 30xy + 36y^2)$
29) $(7x - 10y)(49x^2 + 70xy + 100y^2)$ 30) $b(-2a - b)(4a^2 - 2ab + b^2)$
31) $b^2a(-7a - 3b)(49a^2 - 21ab + 9b^2)$ 32) $5yx(6x - 5y)(36x^2 + 30xy + 25y^2)$

Practice Factoring GCF More Than One Term Solutions

1) $(x + 3)(x + 6)$ 2) $(x + 2)(x + 3)$ 3) $(x - 5)(x - 7)$ 4) $(x - 9)(x - 5)$
5) $(5x + 7)(3x + 4)$ 6) $(4x - 9)(7x - 3)$ 7) $(9x + 5)(3x - 7)$ 8) $(4x - 7)(9x - 2)$
9) $(x - 4)(2x + 1)$ 10) $(x - 7)(x + 1)$ 11) $(x + 2)(4x + 1)$ 12) $(3x - 7)(4x + 9)$
13) $(2x - 9)(5x + 3)$ 14) $(x + 7)(9x + 1)$ 15) $(2x + 3)(6x - 1)$ 16) $(3x + 2)(4x - 1)$
17) $(7x - 5)(4x - 1)$ 18) $(5x - 2)(3x - 1)$ 19) $(5x - 2)(3x + 2)$ 20) $(4x - 7)(9x + 2)$
21) $(2x + 3)(5x + 3)$ 22) $(4x + 1)(7x + 4)$ 23) $(5x - 2)(x^2 + 3x - 2)$
24) $(3x - 5)(2x^2 - 7x + 4)$ 25) $(2x + 7)(5x^2 - 4x + 1)$ 26) $(7x - 3)(3x^2 + x + 1)$
27) $(4x - 5)(7x^2 + x - 1)$ 28) $(5x - 3)(9x^2 + 2x - 1)$ 29) $(x^2 + 2x + 3)(9x - 4)$
30) $(x^2 - 3x + 5)(4x - 7)$ 31) $(x^2 - 7x + 9)(5x + 1)$ 32) $(x^2 - 4x + 7)(9x - 1)$

Practice Factoring Trinomials of the Form ax^2+bx+c

1) $(7b - 6)(b + 9)$ 2) $(5n + 8)(n - 3)$ 3) $(2m - 5)(m - 7)$ 4) $(3k + 2)(k + 4)$
5) $(2x + 7y)(5x + 9y)$ 6) $(2x - 3y)(4x + 5y)$ 7) $(x - y)(4x - 9y)$ 8) $(2x + y)(3x + 10y)$
9) $(5x - 3y)(x + 8y)$ 10) $(7a + 6b)(a - 4b)$ 11) $(2x - 7y)(x - 7y)$ 12) $(3x + 5y)(x + y)$
13) $5a^2(11a - 14)(a - 6)$ 14) $6p(11p + 2)(p + 7)$ 15) $3x^2(13x + 7)(x - 9)$
16) Not factorable   17) $y(3x^2 + 10xy + 70y^2)$ 18) $(2x - 3y)(x + 4y)$
19) $(5x + 3y)(x - 5y)$ 20) $(2u - 3v)(u - 13v)$ 21) Not factorable
22) $4b^2(a - 2b)(9a - 5b)$ 23) $(a + 10b)(14a + 9b)$ 24) $(m - 12n)(9m + 10n)$
25) $-n^2(2n - 3)(6n - 1)$ 26) $3x^2(x - 4)(8x - 9)$ 27) $-8p(p - 8)(9p - 14)$
28) $-(7n - 10)(2n + 9)$ 29) $2(2a + 13b)(7a + 9b)$ 30) $8k(2x + 13y)(5x - 6y)$
31) $2(3x - 4y)(3x - 20y)$ 32) $(u - 8v)(6u - 19v)$

Solutions

Practice Factoring All Mixed-Up Solutions

1) $6(7x + 10)(7x - 10)$ 2) $8(7b^2 + 6)(4b - 5)$ 3) $(3m + 7)^2$ 4) $(x - 9y)(x - 4y)$

5) $(4p + 3q)(7u + 6v)$ 6) Not factorable   7) $(11x + 7y)(x - 4y)$ 8) $(6x + 7y)(2x - 9y)$

9) $(4u + 3)(16u^2 - 12u + 9)$ 10) $y(4x - 7y)(16x^2 + 28xy + 49y^2)$

11) $6(8a - 7m)(7b - 5m)$ 12) $2(x + 6y)(x - 6y)$ 13) Not factorable

14) $4v(u + 3v)(u^2 - 3uv + 9v^2)$ 15) $4y(9x + 8y)^2$ 16) $(b - 4)(9b + 2)$

17) $(2x - 3y)(4x^2 + 6xy + 9y^2)$ 18) $5u(u - 9v)(9u - 5v)$ 19) $12x(10x + 13y)(10x - 13y)$

20) $(4x + 3y)(16x^2 - 12xy + 9y^2)$ 21) $2ab^5(6a + 7b)(3z + 7h)$

22) $(10x + 3y)^2$ 23) $4(49u^2 + 100v^2)$ 24) $4y(4x - 3y)(16x^2 + 12xy + 9y^2)$

25) $(x - 5y)(8x - y)$ 26) $(3u - 2)(9u^2 + 6u + 4)$ 27) $4h(6m^2 - n)(3z + 4h)$

28) $x(x + 5y)(x + 3y)$ 29) $(4m + 3)(16m^2 - 12m + 9)$ 30) $-8n(2a - 7)(7a + 4)$

31) $(3u - 4)(8v - 3u)$ 32) Not factorable   33) $4y(x + 12y)(x + 8y)$

34) $(5m + 7n)^2$ 35) $3b(a - 4b)(a^2 + 4ab + 16b^2)$ 36) $n(11n^2 - 6n - 42)$

37) $(7x - 3v)(2y + 1)$ 38) $(3x + 7y)(9x^2 - 21xy + 49y^2)$ 39) $(2a + 3b)(7a + 6b)$

40) $5(2x + 19y)(2x - 19y)$

Practice Factoring Binomials Higher Degree Solutions

1) $(x^2 + 1)(x - 1)(x + 1)$ 2) $(m^2 + 9)(m - 3)(m + 3)$ 3) $(b^2 + 4)(b - 2)(b + 2)$

4) Not factorable   5) $(x^2 + 16)(x + 4)(x - 4)$ 6) $(3k^2 + 5)(3k^2 - 5)$

7) $(x^3 + 4)(x^3 - 4)$ 8) $(r^3 + 2)(r^3 - 2)$ 9) $(2x^3 + 5)(2x^3 - 5)$ 10) $(3x^3 + 5)(3x^3 - 5)$

11) $(5u^2 + 2)(25u^4 - 10u^2 + 4)$ 12) $(3u^2 + 2)(9u^4 - 6u^2 + 4)$ 13) $(4x^2 - 5)(16x^4 + 20x^2 + 25)$

14) $(2x^2 - 5)(4x^4 + 10x^2 + 25)$ 15) $(u - 1)(u + 1)(u^2 + u + 1)(u^2 - u + 1)$

16) $(x - 2)(x + 2)(x^2 + 2x + 4)(x^2 - 2x + 4)$ 17) $(x^2 - 5)(x^4 + 5x^2 + 25)$

18) $(u^2 + 3)(u^2 + 3u + 3)(u^2 - 3u + 3)$ 19) $(x^4 + 1)(x^2 + 1)(x + 1)(x - 1)$

20) Not factorable   21) $(x - 1)(x^2 + x + 1)(x^6 + x^3 + 1)$

22) $(x + 1)(x^2 - x + 1)(x^6 - x^3 + 1)$

23) $(x - 1)(x^4 + x^3 + x^2 + x + 1)(x + 1)(x^4 - x^3 + x^2 - x + 1)$

24) $(x - 1)(x^2 + x + 1)(x + 1)(x^2 - x + 1)(x^2 + 1)(x^4 - x^2 + 1)$

Practice Factoring Quadratic Form Solutions

1) $(u-1)(u+1)(u-2)(u+2)$ 2) $(m^2-5)^2$ 3) $(u^2-3)(u-2)(u+2)$

4) $(x^2+12)(x^2-8)$ 5) $(x^2-14y^2)(x^2+10y^2)$ 6) $(x-2y)(x+2y)(x^2+9y^2)$

7) $(x^2+10y^2)(x^2+2y^2)$ 8) $(x^2-13y^2)(x^2-11y^2)$ 9) $x(x^2-8y^2)(x^2-7y^2)$

10) $4y(x^2-8y^2)(x^2-5y^2)$ 11) Not factorable   12) $(x^2+6y^2)(x^2+10y^2)$

13) $y(3x^2-13y^2)(x^2+13y^2)$ 14) $(11x^2+14y^2)(x-2y)(x+2y)$

15) $8x(3x^2-2y^2)(x^2+9y^2)$ 16) $(7x^2+3y^2)(x^2-2y^2)$ 17) Not factorable

18) $(2x-3)(2x+3)(4x^2+5)$ 19) $6x(3x^2-2)(5x^2-14)$ 20) $x(4x^2-7)(6x^2-5)$

21) $(x^3+14)(x+2)(x^2-2x+4)$ 22) $(m^4+14)(m^4+5)$

23) $(x^4-6)(x^2-2)(x^2+2)$ 24) $(x-2)(x^2+2x+4)(x+1)(x^2-x+1)$

25) $(3x^4+13y^4)(x-y)(x+y)(x^2+y^2)$ 26) $x^2y^2(13x^4+3y^4)(x^4+y^4)$

27) $(7x^3+13y^3)(x^3+13y^3)$ 28) Not factorable   29) $9y(x^2+y^2)(x^4-x^2y^2+y^4)$

30) $(2x^4-7y^4)(2x^2-y^2)(2x^2+y^2)$ 31) Not factorable

32) $8n(7m^4-9n^4)(2m^2-3n^2)(2m^2+3n^2)$

Practice Solving Quadratic Equations by Factoring Solutions

1) $\left\{\dfrac{5}{12}, 12\right\}$ 2) $\{2, 0\}$ 3) $\{5, -10\}$ 4) $\{-12, -11\}$

5) $\{-4, -2\}$ 6) $\{9, 12\}$ 7) $\{-6, 0\}$ 8) $\{5, 1\}$

9) $\{-6, 8\}$ 10) $\{-5, -4\}$ 11) $\{-7, -12\}$ 12) $\{-9, -6\}$

13) $\{3, -12\}$ 14) $\{8, 6\}$ 15) $\{-3, -10\}$ 16) $\{5, -12\}$

17) $\{-3, 5\}$ 18) $\{6, -1\}$ 19) $\{3, 5\}$ 20) $\{4, 0\}$

21) $\left\{\dfrac{7}{5}, -4\right\}$ 22) $\left\{\dfrac{7}{12}, 0\right\}$ 23) $\left\{\dfrac{6}{11}, \dfrac{5}{7}\right\}$ 24) $\left\{-\dfrac{4}{7}, 1\right\}$

25) $\left\{\dfrac{12}{7}, -1\right\}$ 26) $\left\{-\dfrac{12}{11}, 1\right\}$ 27) $\left\{\dfrac{7}{5}, -7\right\}$ 28) $\left\{-\dfrac{7}{9}, 7\right\}$

29) $\left\{-\dfrac{5}{7}, -4\right\}$ 30) $\left\{\dfrac{7}{2}, -8\right\}$ 31) $\left\{-\dfrac{12}{7}, -5\right\}$ 32) $\left\{-\dfrac{12}{5}, \dfrac{5}{2}\right\}$

Solutions

Practice Simplifying Radicals Solutions

1) $8\sqrt{7}$

2) $4\sqrt{2}$

3) $5\sqrt{3}$

4) $8\sqrt{2}$

5) 14

6) $6\sqrt{3}$

7) $5\sqrt{7}$

8) $3\sqrt{5}$

9) $7\sqrt{6}$

10) $7\sqrt{7}$

11) $18\sqrt{2}$

12) $24\sqrt{2}$

13) $-8\sqrt{6}$

14) $24\sqrt{7}$

15) $28\sqrt{6}$

16) $-24\sqrt{7}$

17) $-16\sqrt{7}$

18) $32\sqrt{2}$

19) $2\sqrt[6]{3}$

20) $4\sqrt[3]{7}$

21) $2\sqrt[3]{4}$

22) $5\sqrt[3]{6}$

23) $2\sqrt[3]{2}$

24) $2\sqrt[3]{6}$

25) $-3\sqrt[3]{5}$

26) -8

27) $3\sqrt[3]{3}$

28) $2\sqrt[4]{6}$

29) $13\sqrt{15}$

30) $15\sqrt{3}$

31) $7\sqrt{13}$

32) 45

Practice Complex Numbers Solutions

1)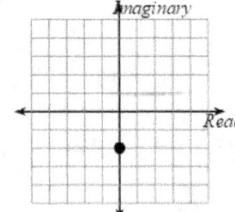

2)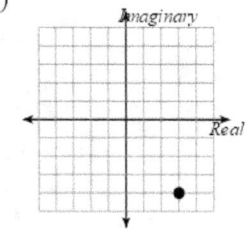

3)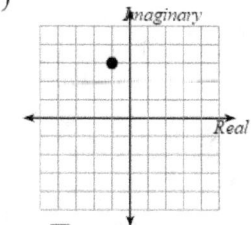

4)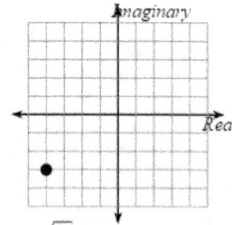

5) 1

6) 5

7) $\sqrt{5}$

8) $2\sqrt{2}$

9) $9i$

10) $6i$

11) $4i\sqrt{6}$

12) $2i\sqrt{5}$

13) $12i\sqrt{5}$

14) $3i\sqrt{2}$

15) $5i\sqrt{3}$

16) $8i\sqrt{3}$

17) $12i\sqrt{3}$

18) $-28i\sqrt{3}$

19) $-35\sqrt{2}$

20) $50i$

21) i

22) -1

23) $-i$

24) 1

25) 1

26) -1

27) $-i$

28) i

29) $2 + 10i$

30) $11 + 7i$

31) $-3 - 9i$

32) $5 + 5i$

33) $-38 + 43i$

34) $-71 + 19i$

35) $-1 + 21i$

36) $10 + 5i$

37) $16 + 30i$

38) $35 + 12i$

39) $-50i$

40) $32i$

41) $\dfrac{9i + 6}{2}$

42) $-5i + 5$

43) $\dfrac{16 - 3i}{20}$

44) $\dfrac{12i + 6}{25}$

45) $\dfrac{-4i + 2}{5}$

46) $\dfrac{-25 - 8i}{13}$

47) $\dfrac{1 + 3i}{5}$

48) $\dfrac{8 - 6i}{5}$

Practice Solving Quadratics by Extracting Square Roots Solutions

1) $\{7, -7\}$ 2) $\{8, -8\}$ 3) $\{7\sqrt{3}, -7\sqrt{3}\}$ 4) $\{3\sqrt{15}, -3\sqrt{15}\}$

5) $\left\{\dfrac{4}{3}, -\dfrac{4}{3}\right\}$ 6) $\left\{\dfrac{3}{7}, -\dfrac{3}{7}\right\}$ 7) $\{6\sqrt{5}, -6\sqrt{5}\}$ 8) $\{3\sqrt{7}, -3\sqrt{7}\}$

9) $\{1, -1\}$ 10) $\left\{\dfrac{5}{7}, -\dfrac{5}{7}\right\}$ 11) $\{2\sqrt{13}, -2\sqrt{13}\}$ 12) $\{\sqrt{34}, -\sqrt{34}\}$

13) $5, -13$ 14) $10, -2$ 15) $-11, -1$ 16) $-1, 13$

17) $3 + 2\sqrt{10}, 3 - 2\sqrt{10}$ 18) $-5 + 6\sqrt{2}, -5 - 6\sqrt{2}$ 19) $4 + 2\sqrt{6}, 4 - 2\sqrt{6}$

20) $-6 + 5\sqrt{3}, -6 - 5\sqrt{3}$ 21) $-4 + 9i, -4 - 9i$ 22) $5 + 8i, 5 - 8i$

23) $6 + 4\sqrt{2}i, 6 - 4\sqrt{2}i$ 24) $-5 + 6\sqrt{2}i, -5 - 6\sqrt{2}i$

Practice Solving Quadratics by Completing the Square Solutions

1) $25; (y + 5)^2$ 2) $9; (x + 3)^2$ 3) $361; (x + 19)^2$ 4) $324; (x - 18)^2$

5) $169; (x - 13)^2$ 6) $144; (y - 12)^2$ 7) $196; (n - 14)^2$ 8) $25; (n - 5)^2$

9) $\dfrac{121}{4}; \left(n - \dfrac{11}{2}\right)^2$ 10) $\dfrac{81}{4}; \left(x - \dfrac{9}{2}\right)^2$ 11) $\dfrac{1}{4}; \left(x - \dfrac{1}{2}\right)^2$ 12) $\dfrac{25}{4}; \left(n + \dfrac{5}{2}\right)^2$

13) $\{8 + 2\sqrt{14}, 8 - 2\sqrt{14}\}$ 14) $\{-9 + \sqrt{14}, -9 - \sqrt{14}\}$ 15) $\{-1 + 3i\sqrt{6}, -1 - 3i\sqrt{6}\}$

16) $\{9 + 3i\sqrt{2}, 9 - 3i\sqrt{2}\}$ 17) $\left\{\dfrac{11 + \sqrt{21}}{2}, \dfrac{11 - \sqrt{21}}{2}\right\}$ 18) $\left\{\dfrac{-7 + 3\sqrt{17}}{2}, \dfrac{-7 - 3\sqrt{17}}{2}\right\}$

19) $\{-8 + 4\sqrt{6}, -8 - 4\sqrt{6}\}$ 20) $\{5 + i\sqrt{23}, 5 - i\sqrt{23}\}$ 21) $\{14, 2\}$

22) $\{-3 + 9i, -3 - 9i\}$ 23) $\{-6 + \sqrt{29}, -6 - \sqrt{29}\}$ 24) $\{7 + \sqrt{10}, 7 - \sqrt{10}\}$

25) $\{8, 2\}$ 26) $\{2, -4\}$ 27) $\{-1 + \sqrt{14}, -1 - \sqrt{14}\}$

28) $\{2 + \sqrt{6}, 2 - \sqrt{6}\}$ 29) $\left\{\dfrac{-3 + 7\sqrt{5}}{2}, \dfrac{-3 - 7\sqrt{5}}{2}\right\}$ 30) $\left\{\dfrac{9 + \sqrt{29}}{2}, \dfrac{9 - \sqrt{29}}{2}\right\}$

31) $\{6, -11\}$ 32) $\left\{\dfrac{-1 + 3\sqrt{21}}{2}, \dfrac{-1 - 3\sqrt{21}}{2}\right\}$ 33) $\{-2 + i\sqrt{13}, -2 - i\sqrt{13}\}$

34) $\left\{\dfrac{3 + 2\sqrt{19}}{3}, \dfrac{3 - 2\sqrt{19}}{3}\right\}$ 35) $\left\{\dfrac{13}{2}, -\dfrac{9}{2}\right\}$ 36) $\left\{\dfrac{-5 + 4i\sqrt{15}}{5}, \dfrac{-5 - 4i\sqrt{15}}{5}\right\}$

37) $\left\{\dfrac{9 + 3i\sqrt{43}}{26}, \dfrac{9 - 3i\sqrt{43}}{26}\right\}$ 38) $\left\{\dfrac{27}{7}, -4\right\}$ 39) $\left\{\dfrac{-9 + \sqrt{53}}{2}, \dfrac{-9 - \sqrt{53}}{2}\right\}$

40) $\left\{\dfrac{-7 + 5\sqrt{29}}{2}, \dfrac{-7 - 5\sqrt{29}}{2}\right\}$

Solutions

Practice Solving Quadratics by Quadratic Formula Solutions

1) $\{-4, 7\}$

2) $\{-6, 3\}$

3) $\left\{ \dfrac{i\sqrt{14}}{4}, -\dfrac{i\sqrt{14}}{4} \right\}$

4) $\{2, -2\}$

5) $\left\{ \dfrac{2 + 2\sqrt{10}}{3}, \dfrac{2 - 2\sqrt{10}}{3} \right\}$

6) $\left\{ \dfrac{-5 + 3\sqrt{15}}{5}, \dfrac{-5 - 3\sqrt{15}}{5} \right\}$

7) $\left\{ \dfrac{-5 + 3i\sqrt{7}}{8}, \dfrac{-5 - 3i\sqrt{7}}{8} \right\}$

8) $\left\{ \dfrac{5 + 2i\sqrt{13}}{11}, \dfrac{5 - 2i\sqrt{13}}{11} \right\}$

9) $\left\{ \dfrac{-1 - 2\sqrt{29}}{5}, \dfrac{-1 + 2\sqrt{29}}{5} \right\}$

10) $\left\{ \dfrac{1 + i\sqrt{21}}{11}, \dfrac{1 - i\sqrt{21}}{11} \right\}$

11) $\left\{ \dfrac{-11 - i\sqrt{131}}{14}, \dfrac{-11 + i\sqrt{131}}{14} \right\}$

12) $\left\{ 5, -\dfrac{19}{4} \right\}$

13) $\left\{ \dfrac{-4 - 3\sqrt{15}}{7}, \dfrac{-4 + 3\sqrt{15}}{7} \right\}$

14) $\left\{ \dfrac{-5 - i\sqrt{3}}{7}, \dfrac{-5 + i\sqrt{3}}{7} \right\}$

15) $\left\{ \dfrac{1 + 2i}{5}, \dfrac{1 - 2i}{5} \right\}$

16) $\left\{ \dfrac{1 + 4\sqrt{6}}{5}, \dfrac{1 - 4\sqrt{6}}{5} \right\}$

17) -155; two imaginary solutions

18) -127; two imaginary solutions

19) 9; two rational solutions

20) -95; two imaginary solutions

21) 0; one rational solution

22) 16; two rational solutions

23) -135; two imaginary solutions

24) 81; two rational solutions

25) 64; two rational solutions

26) 0; one rational solution

27) 169; two rational solutions

28) -155; two imaginary solutions

29) B

30) C

31) A

32) B

33) C

34) A

35) C

36) A

Practice Graphing Quadratics Solutions

1)

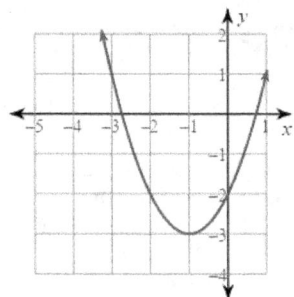

2)

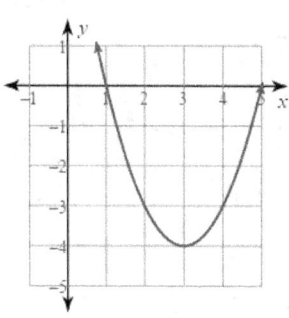

3)

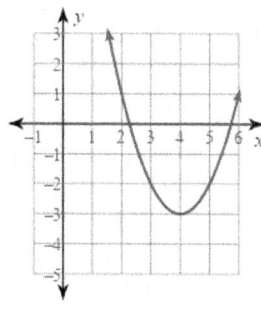

4)

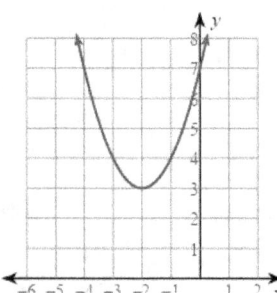

5)

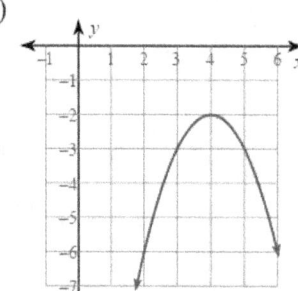

6)

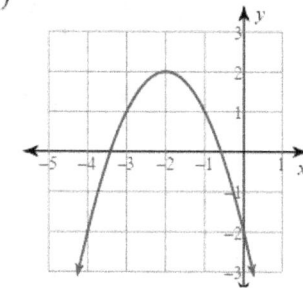

7)

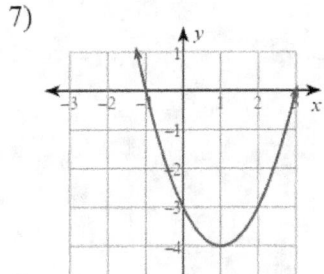

8)

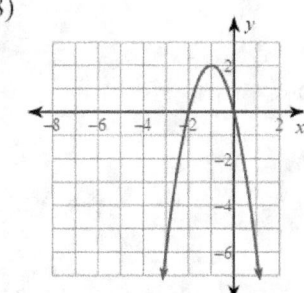

9)

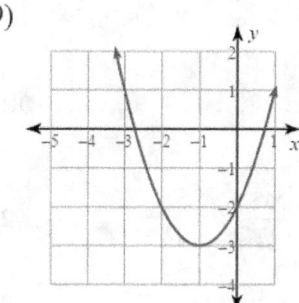

10)

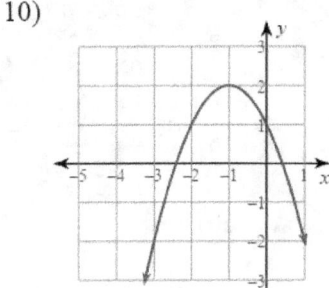

11)

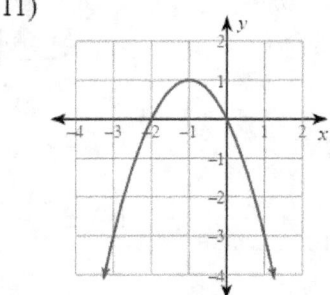

12)

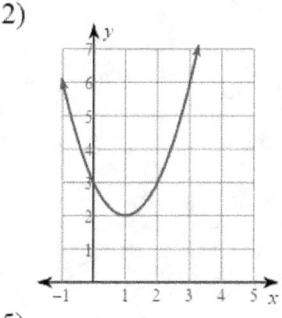

13)

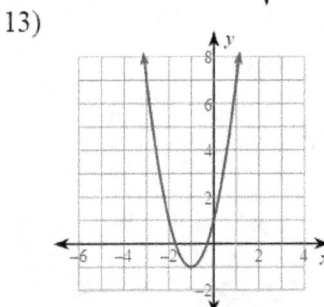

14)

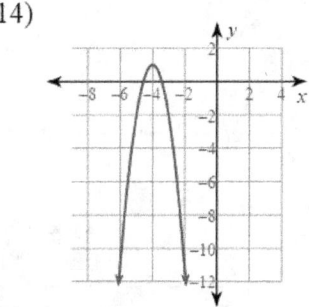

15)

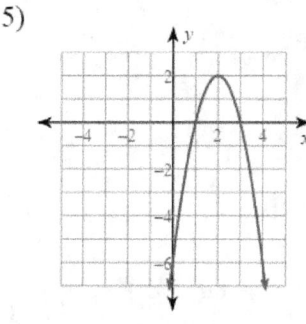

16)

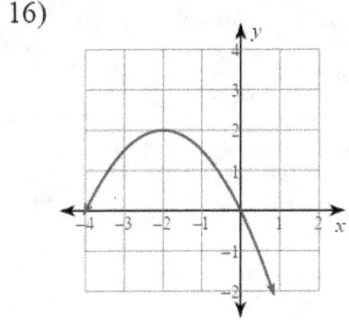

17)

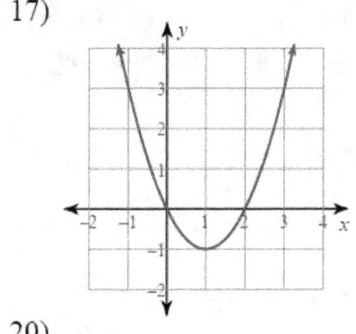

18)

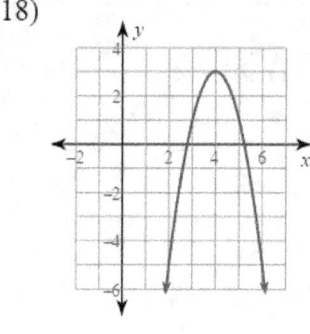

19)

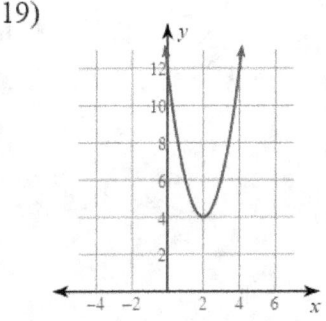

20)

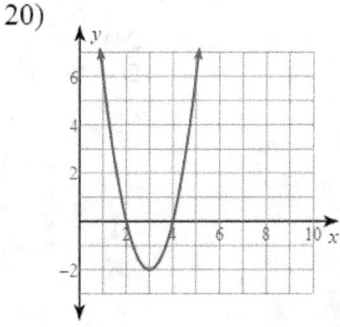

Practice Dividing Polynomials by a Monomial Solutions

1) $\dfrac{5x}{6} + 3 + \dfrac{1}{x}$

2) $1 + \dfrac{5}{r} + \dfrac{1}{5r^2}$

3) $5 + \dfrac{1}{x} + \dfrac{1}{6x^2}$

4) $x + 3 + \dfrac{2}{x}$

5) $x + \dfrac{1}{4} + \dfrac{2}{x}$

6) $\dfrac{b}{2} + \dfrac{1}{2} + \dfrac{1}{2b}$

7) $\dfrac{1}{4} + \dfrac{5}{8k} + \dfrac{1}{k^2}$

8) $\dfrac{1}{3} + \dfrac{4}{9b} + \dfrac{1}{b^2}$

9) $\dfrac{n}{3} + \dfrac{1}{3} + \dfrac{1}{n}$

10) $2a^2 + a + \dfrac{1}{2}$

11) $\dfrac{5x}{8} + \dfrac{1}{4} + \dfrac{1}{x}$

12) $a^2 + a + 2$

13) $2x + 3 + \dfrac{1}{x}$

14) $4m + 4 + \dfrac{1}{3m}$

15) $2p^3 + p^2 + \dfrac{2p}{3}$

16) $\dfrac{a^3}{2} + \dfrac{a^2}{2} + 5a$

17) $\dfrac{5p}{8} + \dfrac{1}{4} + \dfrac{2}{p}$

18) $\dfrac{m}{2} + \dfrac{1}{10} + \dfrac{1}{2m}$

19) $5r^3 + \dfrac{r^2}{3} + \dfrac{r}{3}$

20) $2n^3 + \dfrac{5n^2}{4} + \dfrac{n}{4}$

Practice Dividing Polynomials by Long Division Solutions

1) $n^2 + 10n - 3 + \dfrac{10}{3n + 6}$

2) $k^2 - 10k - 2 + \dfrac{3}{9k - 8}$

3) $7x^2 - 8x + 10 - \dfrac{7}{5x + 9}$

4) $4x^2 - 4x + 4 - \dfrac{7}{4x - 10}$

5) $r^3 - 6r^2 - 7r + 2 - \dfrac{7}{8r + 2}$

6) $m^3 - 6m^2 - 3m + 6 + \dfrac{6}{7m - 8}$

7) $a^3 + 8a^2 - a + 9 - \dfrac{5}{4a + 6}$

8) $x^3 - x^2 - x - 5 - \dfrac{4}{4x + 3}$

9) $x^4 - 4x^3 + 10x^2 - 4x - 8 - \dfrac{1}{3x + 7}$

10) $n^4 - 9n^3 + 4n^2 - n + 5 + \dfrac{8}{6n + 9}$

11) $m^4 + m^3 + m^2 - 5m + 9 - \dfrac{5}{3m + 4}$

12) $n^4 + n^3 - 5n^2 - n + 7 - \dfrac{2}{8n + 1}$

13) $10x^2 - 1 - \dfrac{1}{x - 1}$

14) $9a^2 - 6a + 9 + \dfrac{2}{8a + 1}$

15) $9m + 10 - \dfrac{2}{7m + 8}$

16) $b^3 + 2b^2 - 4b + 8 + \dfrac{5}{3b + 9}$

17) $x^4 + x^3 - \dfrac{2}{2x - 5}$

18) $x^3 + 10x^2 + 6x - 6 + \dfrac{1}{2x - 1}$

19) $n^2 + 2n + 5 - \dfrac{4}{5n + 6}$

20) $p^2 + 3p - 3 - \dfrac{4}{-2 + 5p}$

21) No

22) Yes

23) Yes

24) No

25) No

26) Yes

27) Yes

28) Yes

29) Yes

30) No

31) No

32) No

Practice Dividing Polynomials by Synthetic Division

1) $x^3 + 4x^2 - 10x - 5 - \dfrac{8}{x+1}$

2) $4x^2 - 3x - 7 - \dfrac{3}{x-7}$

3) $7m^3 + 6m^2 + 7m - 2 + \dfrac{8}{m-3}$

4) $a^2 - a + 10 + \dfrac{3}{a-1}$

5) $x^2 + 5x + 4 - \dfrac{6}{x-4}$

6) $k^4 + 3k^3 + 9k^2 - 7k + 7 + \dfrac{10}{k-7}$

7) $n^3 - 5n^2 - 6n + 2 + \dfrac{9}{n-2}$

8) $n^2 - 6n + 5 - \dfrac{8}{n+8}$

9) $v^3 - 2v^2 - 7v - 8 - \dfrac{1}{v-7}$

10) $r^3 - 5r^2 - 5r - 7 - \dfrac{5}{r+4}$

11) $5a^3 + 7a^2 + a - 7 + \dfrac{1}{a+9}$

12) $9p^2 + \dfrac{1}{p+5}$

13) $6a^3 + 10a^2 - 7a - 6 - \dfrac{5}{a+2}$

14) $r^3 + 9 + \dfrac{9}{r-4}$

15) $x^3 + 8x^2 - 6x + 2 - \dfrac{8}{x-8}$

16) $r^2 - 10r + 4 + \dfrac{2}{r+6}$

17) $3v^3 + 7v^2 + 9v - 3 - \dfrac{1}{3v+2}$

18) $k^3 - 2k^2 - k - 8 - \dfrac{5}{4k-1}$

19) $6x^3 - 10x^2 + 2x + 10 - \dfrac{5}{2x-8}$

20) $4n^3 - 2n^2 + 4n + 5 - \dfrac{3}{7n+6}$

21) $v^4 + 9v^2 - v - 5 + \dfrac{1}{9v+1}$

22) $x^2 + \dfrac{7}{6x+4}$

23) $a^4 + 5a - 8 - \dfrac{9}{4a-6}$

24) $n^2 - 9n - 3 + \dfrac{6}{2n+5}$

Practice Graphs of Polynomials Solutions

1) 5 quintic negative   2) 4 quartic positive   3) 3 cubic negative   4) 0 constant positive

5) 1 linear positive   6) 5 quintic positive   7) 2 quadratic positive   8) 2 quadratic negative

9) 4 quartic positive   10) 3 cubic positive   11) 1 linear positive   12) 4 quartic negative

13) 5 quintic positive   14) 2 quadratic positive   15) 3 cubic positive

16) 0 constant negative

Practice Zeros, Roots, and Factors Solutions

1) $0, 4, -5, \dfrac{4}{3}$

2) $-4, 5, -\dfrac{3}{4}$

3) $(3, 0), (-5, 0), \left(-\dfrac{7}{2}, 0\right)$

4) $(0, 0), (-3, 0), (2, 0), \left(\dfrac{1}{2}, 0\right)$

5) $x(x+7)(x-2)(5x+6)$

6) $(x+2)(2x+3)(5x-2)(x-6)$

7) Zeros: $-4, 2$
Factors:
$(x-2)(x+4)$

8) Zeros: $-1, 3$
Factors: $-(x-3)(x+1)$

9) Zeros: -2
Factors: $(x+2)(x+2)$

10) Zeros: $-1, 2$
Factors: $-(x+1)(x+1)(x-2)$

11) Zeros: $1, -2, 4$
Factors: $(x-1)(x+2)(x-4)$

12) Zeros: $0, 2, 5$
Factors: $-x(x-2)(x-5)$

13) Zeros: $-4, 2$
Factors: $-0.2(x-2)(x-2)(x+4)$

14) Zeros: $-1, 5$
Factors: $-0.2(x+1)(x+1)(x-5)$

15) Zeros: $-1, 0, 2, 4$
Factors: $x(x-2)(x+1)(x-4)$

16) Zeros: −2, −1, 1, 3
   Factors: $-(x-1)(x+2)(x+1)(x-3)$

17) Zeros: −1, 2, 4
   Factors: $(x+1)(x-2)(x-4)$

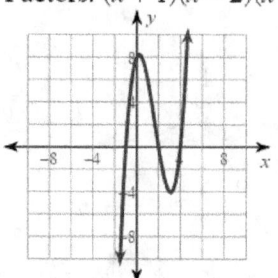

18) Zeros: −3, 1, 2
   Factors: $(x-1)(x-2)(x+3)$

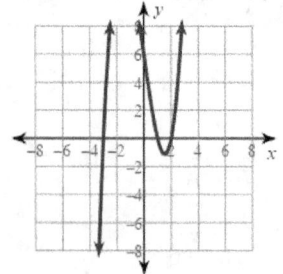

19) Zeros: −2, −1, 1, 4
   Factors: $(x-1)(x+1)(x-4)(x+2)$

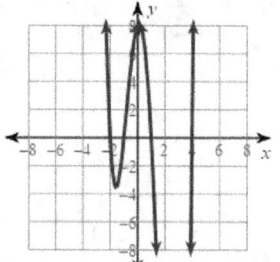

20) Zeros: −4, −3, −2, 2
   Factors: $(x-2)(x+3)(x+4)(x+2)$

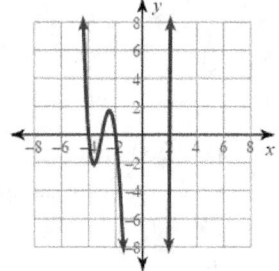

21) Zeros: −4, 0, 2
   Factors: $x(x-2)(x-2)(x+4)$

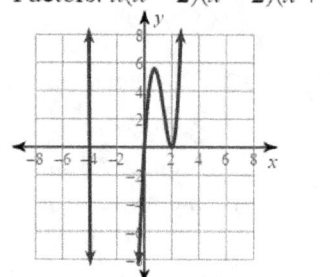

22) Zeros: −1, 0, 5
   Factors: $x(x+1)(x+1)(x-5)$

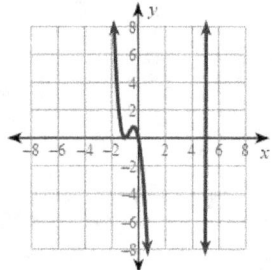

23) Zeros: −1, 0, 1, 2, 3
   Factors: $-x(x-1)(x+1)(x-2)(x-3)$

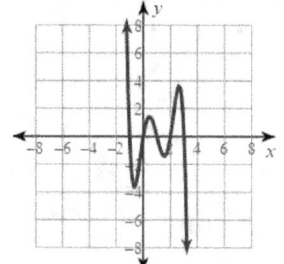

24) Zeros: −2, −1, 0, 2, 3
   Factors: $x(x+1)(x+2)(x-2)(x-3)$

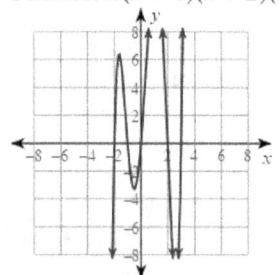

Practice Rational Root Theorem Solutions

1) $\pm 1, \pm 2, \pm 3, \pm 6, \pm \frac{1}{2}, \pm \frac{3}{2}$

2) $\pm 1, \pm 3, \pm 5, \pm 15, \pm \frac{1}{3}, \pm \frac{5}{3}$

3) $\pm 1, \pm 3, \pm 9, \pm 27, \pm \frac{1}{2}, \pm \frac{3}{2}, \pm \frac{9}{2}, \pm \frac{27}{2}$

4) $\pm 1, \pm 2, \pm 4, \pm 5, \pm 8, \pm 10, \pm 20, \pm 40, \pm \frac{1}{2}, \pm \frac{5}{2}$

5) $\pm 1, \pm 3, \pm 9, \pm \frac{1}{5}, \pm \frac{3}{5}, \pm \frac{9}{5}$

6) $\pm 1, \pm 5, \pm 25, \pm \frac{1}{2}, \pm \frac{5}{2}, \pm \frac{25}{2}$

7) $\pm 1, \pm 5, \pm \frac{1}{2}, \pm \frac{5}{2}, \pm \frac{1}{5}, \pm \frac{1}{10}$

8) $\pm 1, \pm 7, \pm \frac{1}{3}, \pm \frac{7}{3}$

9) $\pm 1, \pm 2, \pm 3, \pm 6, \pm 9, \pm 18$

10) $\pm 1, \pm \frac{1}{2}$

11) $\pm 1, \pm \frac{1}{5}$

12) $\pm 1, \pm 2, \pm 4, \pm 8, \pm 16$

13) Possible rational zeros: $\pm 1, \pm \frac{1}{3}$

Rational zeros: $\left\{ -\frac{1}{3}, 1, -1 \right\}$

14) Possible rational zeros: $\pm 1, \pm \frac{1}{2}$

Rational zeros: $\left\{ \frac{1}{2}, -1, 1 \right\}$

15) Possible rational zeros: $\pm 1, \pm \frac{1}{2}$

Rational zeros: $\left\{ -\frac{1}{2}, -1 \text{ mult. } 2 \right\}$

16) Possible rational zeros:

$\pm 1, \pm 5, \pm 25, \pm \frac{1}{3}, \pm \frac{5}{3}, \pm \frac{25}{3}$

Rational zeros: $\{5\}$

17) Possible rational zeros: $\pm 1, \pm 2$

Rational zeros: $\{2\}$

18) Possible rational zeros: $\pm 1, \pm 5, \pm 7, \pm 35$

Rational zeros: $\{5\}$

19) Possible rational zeros: $\pm 1, \pm 5, \pm \frac{1}{5}$

Rational zeros: $\left\{ 1, -5, -\frac{1}{5} \right\}$

20) Possible rational zeros:

$\pm 1, \pm 2, \pm 3, \pm 5, \pm 6, \pm 10, \pm 15, \pm 30$

Rational zeros: $\{-3\}$

Practice Descartes' Rule of Signs Solutions

1) Possible # positive real zeros: 2 or 0
Possible # negative real zeros: 0

2) Possible # positive real zeros: 1
Possible # negative real zeros: 1

3) Possible # positive real zeros: 1
Possible # negative real zeros: 2 or 0

4) Possible # positive real zeros: 2 or 0
Possible # negative real zeros: 1

5) Possible # positive real zeros: 1
Possible # negative real zeros: 1

6) Possible # positive real zeros: 2 or 0
Possible # negative real zeros: 1

7) Possible # positive real zeros: 4, 2, or 0
Possible # negative real zeros: 1

8) Possible # positive real zeros: 1
Possible # negative real zeros: 4, 2, or 0

9) Possible # positive real zeros: 1
Possible # negative real zeros: 4, 2, or 0

10) Possible # positive real zeros: 4, 2, or 0
Possible # negative real zeros: 1

11) Possible # positive real zeros: 1
Possible # negative real zeros: 1

12) Possible # positive real zeros: 2 or 0
Possible # negative real zeros: 3 or 1

13) Possible # positive real zeros: 2 or 0
Possible # negative real zeros: 2 or 0

14) Possible # positive real zeros: 1
Possible # negative real zeros: 1

15) Possible # positive real zeros: 2 or 0
Possible # negative real zeros: 1

16) Possible # positive real zeros: 1
Possible # negative real zeros: 2 or 0

17) Possible # positive real zeros: 2 or 0
Possible # negative real zeros: 1

18) Possible # positive real zeros: 1
Possible # negative real zeros: 2 or 0

19) Possible # positive real zeros: 4, 2, or 0
Possible # negative real zeros: 1

20) Possible # positive real zeros: 3 or 1
Possible # negative real zeros: 0

Solutions

Practice Fundamental Theorem of Algebra Solutions

1) # of complex zeros: 3
   Possible # of real zeros: 3 or 1
   Possible # of imaginary zeros: 2 or 0
   Other Roots: $1 - i\sqrt{2}$

2) # of complex zeros: 3
   Possible # of real zeros: 3 or 1
   Possible # of imaginary zeros: 2 or 0
   Other Roots: $2 - 2i\sqrt{3}$

3) # of complex zeros: 4
   Possible # of real zeros: 4, 2, or 0
   Possible # of imaginary zeros: 4, 2, or 0
   Other Roots: $-\dfrac{i\sqrt{10}}{5}, -i$

4) # of complex zeros: 5
   Possible # of real zeros: 5, 3, or 1
   Possible # of imaginary zeros: 4, 2, or 0
   Other Roots: $\dfrac{i\sqrt{15}}{5}, -i$

5) # of complex zeros: 5
   Possible # of real zeros: 5, 3, or 1
   Possible # of imaginary zeros: 4, 2, or 0
   Other Root: $\dfrac{-5 + i\sqrt{55}}{4}$

6) # of complex zeros: 4
   Possible # of real zeros: 4, 2, or 0
   Possible # of imaginary zeros: 4, 2, or 0
   Other Roots: $2i\sqrt{2}, -\dfrac{i\sqrt{15}}{5}$

7) # of complex zeros: 5
   Possible # of real zeros: 5, 3, or 1
   Possible # of imaginary zeros: 4, 2, or 0
   Other Roots: $1 - \sqrt{47}, i$

8) # of complex zeros: 6
   Possible # of real zeros: 6, 4, 2, or 0
   Possible # of imaginary zeros: 6, 4, 2, or 0
   Other roots: $-1 - 3i, -\sqrt{6}$

9) # of complex zeros: 8
   Possible # of real zeros: 8, 6, 4, 2, or 0
   Possible # of imaginary zeros: 8, 6, 4, 2, or 0
   Other Roots: $-3 - \sqrt{10}, \dfrac{-1 - i\sqrt{3}}{2}$

10) # of complex zeros: 6
   Possible # of real zeros: 6, 4, 2, or 0
   Possible # of imaginary zeros: 6, 4, 2, or 0
   Other roots: $2 - \sqrt{33}, \sqrt{2}, i$

11) # of complex zeros: 11
   Possible # of real zeros: 11, 9, 7, 5, 3, or 1
   Possible # of imaginary zeros: 10, 8, 6, 4, 2, or 0
   Other Roots: $-1 - \sqrt{37}, 1 + i\sqrt{3}, \dfrac{-1 - i\sqrt{3}}{2}$

12) # of complex zeros: 10
   Possible # of real zeros: 10, 8, 6, 4, 2, or 0
   Possible # of imaginary zeros: 10, 8, 6, 4, 2, or 0
   Other Roots: $3 - i, i, -\sqrt{5}, i\sqrt{5}$

Practice Factoring Higher Degrees Solutions

1) $\left\{\sqrt{5}, -\sqrt{5}, i\sqrt{5}, -i\sqrt{5}, \sqrt{2}, -\sqrt{2}, i\sqrt{2}, -i\sqrt{2}\right\}$
2) $\left\{1, -1, i\sqrt{6}, -i\sqrt{6}\right\}$ 3) $\left\{\sqrt{5}\text{ mult. }2, -\sqrt{5}\text{ mult. }2\right\}$ 4) $\left\{0\text{ mult. }2, 4, -2\right\}$
5) $\left\{i\sqrt{7}, -i\sqrt{7}, \sqrt{2}, -\sqrt{2}\right\}$ 6) $\left\{2i\sqrt{2}, -2i\sqrt{2}, i, -i\right\}$
7) $\left\{0\text{ mult. }2, -1 + \sqrt{26}, -1 - \sqrt{26}\right\}$ 8) $\left\{1, -1, i, -i, \sqrt{5}, -\sqrt{5}, i\sqrt{5}, -i\sqrt{5}\right\}$
9) $\left\{1\text{ mult. }2, -1\text{ mult. }2, i\text{ mult. }2, -i\text{ mult. }2\right\}$ 10) $\left\{i\sqrt{2}, -i\sqrt{2}, \sqrt{7}, -\sqrt{7}\right\}$
11) $\left\{3, -3, \sqrt{7}, -\sqrt{7}\right\}$ 12) $\left\{2i, -2i, i\sqrt{2}, -i\sqrt{2}\right\}$

13) Factors to: $(x + 3)(x - 2) = 0$
   Roots: $\{-3, 2\}$

14) Factors to: $x^2 + 4x + 20 = 0$
   Roots: $\{-2 + 4i, -2 - 4i\}$

15) Factors to: $x(x^2 - 4x - 11) = 0$
   Roots: $\{0, 2 + \sqrt{15}, 2 - \sqrt{15}\}$

16) Factors to: $x(x^2 - 2x + 17) = 0$
   Roots: $\{0, 1 + 4i, 1 - 4i\}$

17) Factors to: $(x^2 + 5)(x^2 + 6) = 0$
   Roots: $\{i\sqrt{5}, -i\sqrt{5}, i\sqrt{6}, -i\sqrt{6}\}$

18) Factors to: $(x^2 + 5)(x^2 - 7) = 0$
   Roots: $\{i\sqrt{5}, -i\sqrt{5}, \sqrt{7}, -\sqrt{7}\}$

19) Factors to: $(x^2 - 8)(x^2 + 1) = 0$
   Roots: $\{2\sqrt{2}, -2\sqrt{2}, i, -i\}$

20) Factors to: $x^2 - 8x - 1 = 0$
   Roots: $\{4 + \sqrt{17}, 4 - \sqrt{17}\}$

21) Factors to: $x^2(x + 2)^2(x - 2) = 0$
   Roots: $\{0 \text{ mult. } 2, -2 \text{ mult. } 2, 2\}$

22) Factors to: $x(x - 3)(x + 5) = 0$
   Roots: $\{0, 3, -5\}$

23) Factors to: $x(x^2 - 6x - 28) = 0$
   Roots: $\{0, 3 + \sqrt{37}, 3 - \sqrt{37}\}$

24) Factors to: $(x^2 + 6)(x^2 + 8) = 0$
   Roots: $\{i\sqrt{6}, -i\sqrt{6}, 2i\sqrt{2}, -2i\sqrt{2}\}$

25) Factors to: $(x^2 + 6)(x - 3)(x + 3) = 0$
   Roots: $\{i\sqrt{6}, -i\sqrt{6}, 3, -3\}$

26) Factors to: $(x - 1)(x^2 + x + 1)(x + 1)(x^2 - x + 1) = 0$
   Roots: $\left\{1, \dfrac{-1 + i\sqrt{3}}{2}, \dfrac{-1 - i\sqrt{3}}{2}, -1, \dfrac{1 + i\sqrt{3}}{2}, \dfrac{1 - i\sqrt{3}}{2}\right\}$

27) Factors to: $(x - 2)(x^2 + 5)(x^2 - 3) = 0$
   Roots: $\{2, i\sqrt{5}, -i\sqrt{5}, \sqrt{3}, -\sqrt{3}\}$

28) Factors to: $(x - 2)(x^2 - 2)(x^2 + 8) = 0$
   Roots: $\{2, \sqrt{2}, -\sqrt{2}, 2i\sqrt{2}, -2i\sqrt{2}\}$

29) $\left\{1, -1, \dfrac{2\sqrt{3}}{3}, -\dfrac{2\sqrt{3}}{3}, \dfrac{2i\sqrt{3}}{3}, -\dfrac{2i\sqrt{3}}{3}, 1 + 3i, 1 - 3i\right\}$

30) $\left\{-2, 1 + i\sqrt{3}, 1 - i\sqrt{3}, -\dfrac{1}{2}, \dfrac{1 + i\sqrt{3}}{4}, \dfrac{1 - i\sqrt{3}}{4}, 2 + \sqrt{7}, 2 - \sqrt{7}\right\}$

31) $\left\{1, -1, i, -i, \dfrac{\sqrt{3}}{2}, -\dfrac{\sqrt{3}}{2}, \dfrac{i\sqrt{3}}{2}, -\dfrac{i\sqrt{3}}{2}, 1 + i, 1 - i\right\}$

32) $\left\{\dfrac{1}{2}, \dfrac{-1 + i\sqrt{3}}{4}, \dfrac{-1 - i\sqrt{3}}{4}, 1, \dfrac{-1 + i\sqrt{3}}{2}, \dfrac{-1 - i\sqrt{3}}{2}, -1, \dfrac{1 + i\sqrt{3}}{2}, \dfrac{1 - i\sqrt{3}}{2}, -4 + i, -4 - i\right\}$

Practice Graphing Polynomials Solutions

1)

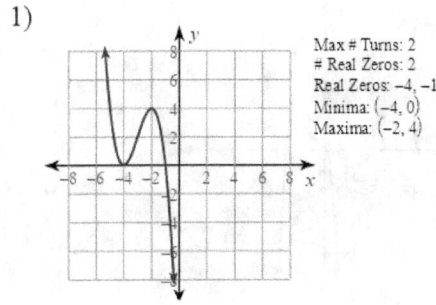

Max # Turns: 2
Real Zeros: 2
Real Zeros: −4, −1
Minima: (−4, 0)
Maxima: (−2, 4)

2)

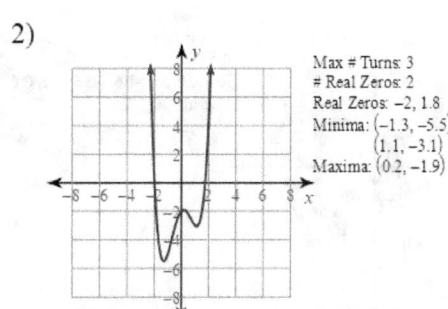

Max # Turns: 3
Real Zeros: 2
Real Zeros: −2, 1.8
Minima: (−1.3, −5.5)
          (1.1, −3.1)
Maxima: (0.2, −1.9)

Solutions

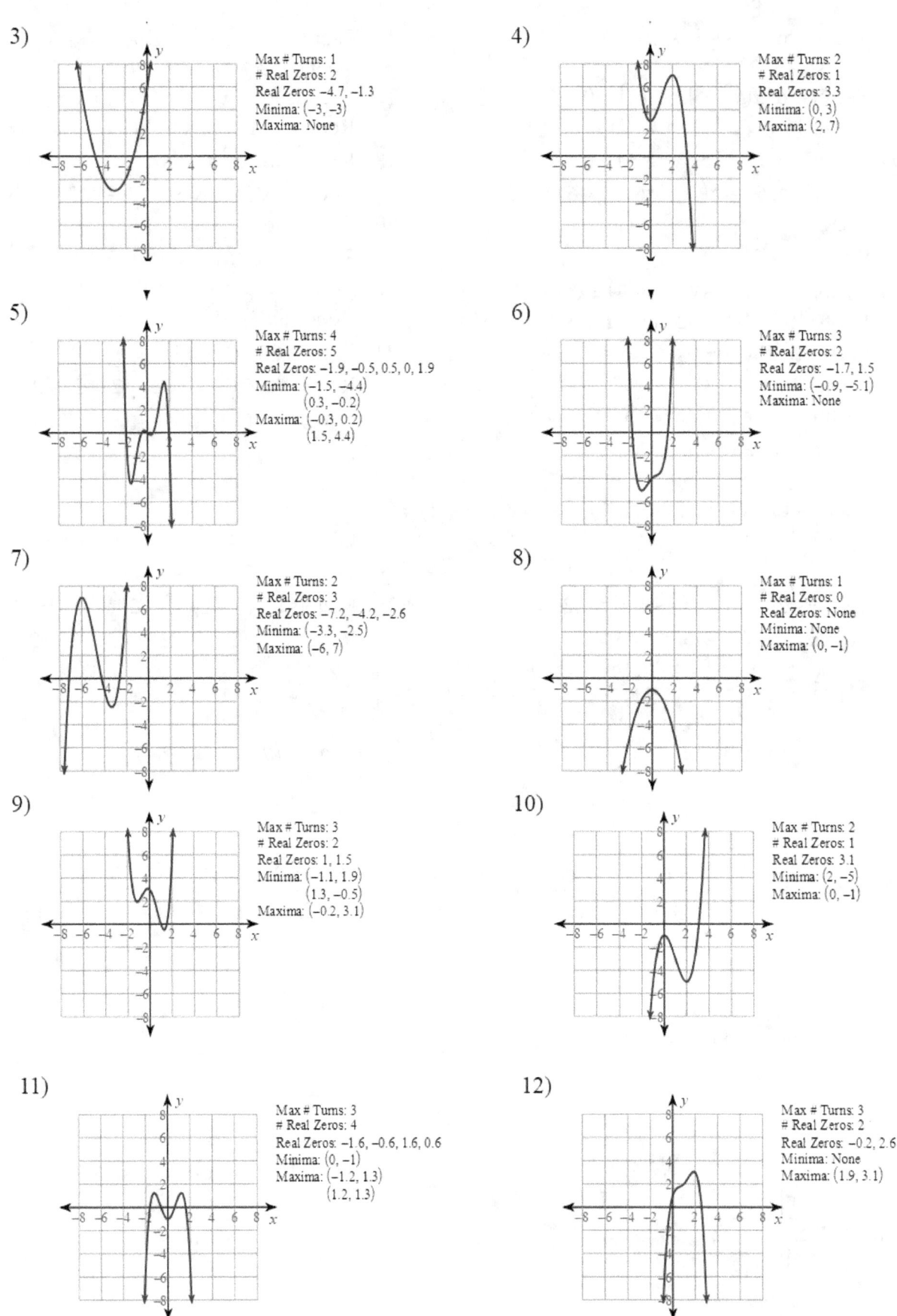

3)

Max # Turns: 1
Real Zeros: 2
Real Zeros: −4.7, −1.3
Minima: (−3, −3)
Maxima: None

4)

Max # Turns: 2
Real Zeros: 1
Real Zeros: 3.3
Minima: (0, 3)
Maxima: (2, 7)

5)

Max # Turns: 4
Real Zeros: 5
Real Zeros: −1.9, −0.5, 0.5, 0, 1.9
Minima: (−1.5, −4.4)
         (0.3, −0.2)
Maxima: (−0.3, 0.2)
         (1.5, 4.4)

6)

Max # Turns: 3
Real Zeros: 2
Real Zeros: −1.7, 1.5
Minima: (−0.9, −5.1)
Maxima: None

7)

Max # Turns: 2
Real Zeros: 3
Real Zeros: −7.2, −4.2, −2.6
Minima: (−3.3, −2.5)
Maxima: (−6, 7)

8)

Max # Turns: 1
Real Zeros: 0
Real Zeros: None
Minima: None
Maxima: (0, −1)

9)

Max # Turns: 3
Real Zeros: 2
Real Zeros: 1, 1.5
Minima: (−1.1, 1.9)
         (1.3, −0.5)
Maxima: (−0.2, 3.1)

10)

Max # Turns: 2
Real Zeros: 1
Real Zeros: 3.1
Minima: (2, −5)
Maxima: (0, −1)

11)

Max # Turns: 3
Real Zeros: 4
Real Zeros: −1.6, −0.6, 1.6, 0.6
Minima: (0, −1)
Maxima: (−1.2, 1.3)
         (1.2, 1.3)

12)

Max # Turns: 3
Real Zeros: 2
Real Zeros: −0.2, 2.6
Minima: None
Maxima: (1.9, 3.1)

13)

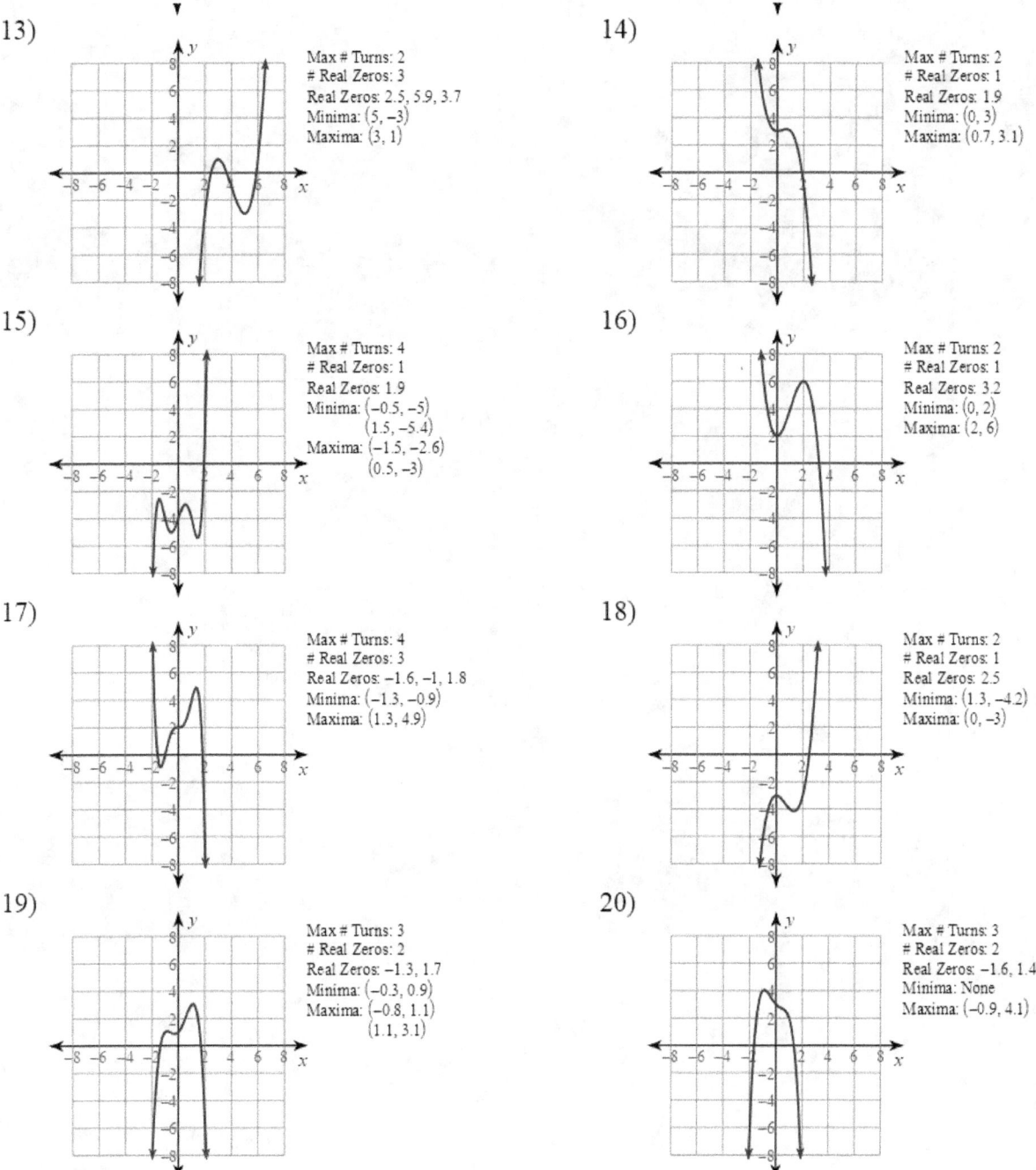

Max # Turns: 2
Real Zeros: 3
Real Zeros: 2.5, 5.9, 3.7
Minima: (5, −3)
Maxima: (3, 1)

14)

Max # Turns: 2
Real Zeros: 1
Real Zeros: 1.9
Minima: (0, 3)
Maxima: (0.7, 3.1)

15)

Max # Turns: 4
Real Zeros: 1
Real Zeros: 1.9
Minima: (−0.5, −5)
         (1.5, −5.4)
Maxima: (−1.5, −2.6)
         (0.5, −3)

16)

Max # Turns: 2
Real Zeros: 1
Real Zeros: 3.2
Minima: (0, 2)
Maxima: (2, 6)

17)

Max # Turns: 4
Real Zeros: 3
Real Zeros: −1.6, −1, 1.8
Minima: (−1.3, −0.9)
Maxima: (1.3, 4.9)

18)

Max # Turns: 2
Real Zeros: 1
Real Zeros: 2.5
Minima: (1.3, −4.2)
Maxima: (0, −3)

19)

Max # Turns: 3
Real Zeros: 2
Real Zeros: −1.3, 1.7
Minima: (−0.3, 0.9)
Maxima: (−0.8, 1.1)
         (1.1, 3.1)

20)

Max # Turns: 3
Real Zeros: 2
Real Zeros: −1.6, 1.4
Minima: None
Maxima: (−0.9, 4.1)